Hormesis

Mark P. Mattson · Edward J. Calabrese
Editors

Hormesis

A Revolution in Biology, Toxicology and Medicine

Springer

Editors
Mark P. Mattson
Laboratory of Neurosciences
GRC 4F01
National Institute on Aging
5600 Nathan Shock Drive
Baltimore, MD 21224
USA
mattsonm@grc.nia.nih.gov

Edward J. Calabrese
Department of Environmental Health Sciences
University of Massachusetts
Northeast Regional Environmental Public
N344 Morrill Science Center
Amherst, MA 01003-5712
USA
edward@schoolph.umass.edu

ISBN 978-1-60761-494-4 e-ISBN 978-1-60761-495-1
DOI 10.1007/978-1-60761-495-1
Springer New York Dordrecht Heidelberg London

Library of Congress Control Number: 2009938828

Printed on acid-free paper

Springer is part of Springer Science+Business Media (www.springer.com)

Preface

The term *hormesis* is defined as "a process in which exposure to a low dose of a chemical agent or environmental factor that is damaging at higher doses induces an adaptive beneficial effect on the cell or organism" (Calabrese et al., 2007; Mattson, 2008). To survive and reproduce in harsh competitive environments, organisms and their cellular components have, through evolution, developed molecular mechanisms to respond adaptively to various hazards or "stressors" that they encounter. Examples of such stressors include chemicals ingested in food and water (metals, phytochemicals, etc.), increased energy expenditure (running, fighting, cognitive challenges, etc.), and reduced energy availability (food scarcity), among others. In most cases, the response of the cell or organism to the stressor exhibits a biphasic dose response, with beneficial/adaptive responses at low doses (improved function, increased resistance to damage and disease) and adverse/destructive effects (dysfunction, molecular damage, or even death) at high doses. The prevalence of the biphasic (hormetic) dose response characteristic of biological systems merits consideration of hormesis as a fundamental principle of biology.

In this book, my colleagues and I present evidence from a range of biological systems that hormesis is indeed at the epicenter of the molecular and cellular responses to their environment. Many of the thousands of examples of hormesis (biphasic dose responses with stimulatory/beneficial effects at low doses and inhibitory/toxic effects at high doses) come from the field of toxicology (Calabrese, 2008), and yet the Environmental Protection Agency (EPA) continues to largely ignore the important scientific fact of the biphasic dose response. Their approach is to reduce the levels of "toxins" in the environment as much as possible. However, it is clear that at least in some cases human health may be adversely affected by removing "toxic" chemicals from the environment. Prominent examples are metals such as selenium, zinc, and iron, all of which are toxic when consumed in high amounts but are essential for health in low amounts (Dodig and Cepelak, 2004; Frassinetti et al., 2006; Wright and Baccarelli, 2007). Other major, emerging examples are phytochemicals that function as insect repellants (toxins) in plants but stimulate adaptive stress response pathways when consumed by humans (Cheng and Mattson, 2006).

Of interest, many endogenous cellular signaling pathways exert their effects on cellular physiology (cell division, the growth of muscle and nerve cells, and even behaviors such as learning and memory) through hormetic mechanisms. For example, the excitatory neurotransmitter glutamate is released from presynaptic terminals at synapses, where it then activates receptors that are coupled to calcium influx into the dendrites of the postsynaptic neuron. In this way glutamate plays a fundamental role in the function of neuronal circuits involved in sensory processing, motor responses, learning and memory, and emotional behaviors. These low levels of glutamate also activate adaptive stress responses that include the production of proteins that help to protect the neurons against more-severe stress. These stress resistance proteins include neurotrophic factors, antioxidant enzymes, and antiapoptotic proteins such as Bcl2. However, abnormally high levels of glutamate resulting from increased release and/or decreased removal at synapses can cause the degeneration and death of neurons. The latter neurotoxic effects of excessive activation of glutamate receptors occur in patients with epilepsy, stroke, traumatic brain and spinal cord injury, and possibly Alzheimer's, Parkinson's, and Huntington's diseases. The situation is similar with other signaling pathways in other tissues and organs. Consequently, the scientific and biomedical professions should work to elucidate the molecular components of hormetic signaling pathways and apply that knowledge to the development of novel hormesis-based preventative and therapeutic interventions for many different human diseases.

This book comprises 10 chapters, with contributions from more than a dozen authors to the writing of one or more of the chapters. The first chapter describes the concept of hormesis, the prevalence of biphasic dose responses in biological systems, and implications of hormesis for the future of science, medicine, and public policy decisions. The second chapter focuses on the role of hormesis in toxicology and risk assessment, with a focus on environmental toxins. A chapter that considers hormesis from an evolutionary perspective provides several examples of how organisms not only developed mechanisms to respond adaptively to "toxins," but also actually incorporated those chemicals into their metabolic systems. The next three chapters describe several of the most highly conserved signaling mechanisms that mediate hormetic responses of cells and organisms exposed to subtoxic doses of chemicals and other stressors. These include G protein–coupled receptors and signaling pathways that lead to the induction of genes that encode cytoprotective proteins such as heat-shock proteins, antioxidant enzymes, and growth factors. The complexity of receptor systems and cellular responses provides a rich venue for understanding the intricacies of the molecular mediators of hormesis. The health benefits of exercise and dietary modification (particularly dietary energy restriction) are well known. Two chapters provide evidence that many of the beneficial effects of exercise and dietary modification result from activation of hormetic signaling pathways in cells throughout the body.

Particularly intriguing are the prominent hormetic effects of exercise and dietary energy restriction on brain health. Data suggest that hormetic mechanisms may be compromised during aging, and such impairments may contribute to the development of a range of age-related diseases. We are in the midst of an epidemic of obesity and diabetes in the United States, and this major health problem is spreading to industrialized countries in all continents. A chapter describes evidence that the "couch potato" lifestyle that causes obesity and diabetes does so, in part, by suppressing the activation of hormetic response pathways. The book concludes with a chapter entitled The Hormetic Pharmacy that considers the role of hormesis-based mechanisms of action in the future of natural products and man-made drugs for disease prevention and treatment. Early in the 16th century, Paracelsus recognized that all drugs are poisonous at high doses and that careful evaluation of dose-response relationships are necessary for optimizing treatments. In this book we emphasize our newer recognition of the great potential of hormesis-based approaches for drug discovery, as well as for the optimization of dietary and lifestyle factors to improve the quality of life.

References

Calabrese EJ et al. (2007) Biological stress response terminology: integrating the concepts of adaptive response and preconditioning stress within a hormetic dose-response framework. Toxicol Appl Pharmacol 222: 122–128.

Calabrese EJ (2008) Hormesis: why it is important to toxicology and toxicologists. Environ Toxicol Chem. 27:1451–1474.

Cheng A, Mattson MP (2006) Neurohormetic phytochemicals: low-dose toxins that induce adaptive neuronal stress responses. Trends Neurosci 29: 632–639.

Dodig S, Cepelak I (2004) The facts and controversies about selenium. Acta Pharm 54: 261–276.

Frassinetti S, Bronzetti G, Caltavuturo L, Cini M, Croce CD (2006) The role of zinc in life: a review. J Environ Pathol Toxicol Oncol 25: 597–610.

Mattson MP (2008) Hormesis defined. Ageing Res Rev 7: 1–7.

Wright RO, Baccarelli A (2007) Metals and neurotoxicology. J Nutr 137: 2809–2813.

Baltimore, Maryland Mark P. Mattson

Contents

About the Editors

Mark P. Mattson, Ph.D., is Chief of the Laboratory of Neurosciences at the National Institute on Aging in Baltimore, where he leads a multifaceted research team that applies cutting-edge technologies in research aimed at understanding molecular and cellular mechanisms of brain aging and the pathogenesis of neurodegenerative disorders. He is also a professor in the Department of Neuroscience at Johns Hopkins University School of Medicine. He has published more than 450 original research articles and numerous review articles and has edited 10 books in the areas of mechanisms of aging and neurodegenerative disorders. Dr. Mattson has trained more than 60 postdoctoral and predoctoral students and is the most highly cited neuroscientist in the world.

Edward J. Calabrese, Ph.D., is a professor and Program Director of Environmental Health Science at the University of Massachusetts in Amherst. His research focuses on environmental toxicology, with an emphasis on biological factors, including genetic and nutritional factors that enhance susceptibility to pollutant toxicity and the environmental implications of toxicological hormesis. Dr. Calabrese has researched extensively in the area of host factors affecting susceptibility to pollutants and is the author of more than 600 papers in scholarly journals, as well as 24 books in the field of toxicology and environmental pollution. Dr. Calabrese has received numerous awards, including, most recently, the prestigious Marie Curie Prize.

Contributors

Mark P. Mattson Laboratory of Neurosciences, National Institute on Aging, Intramural Research Program, Baltimore, MD 21224, USA, mattsonm@grc.nia.nih.gov

Edward J. Calabrese Department of Environmental Health Sciences, School of Public Health and Health Sciences, University of Massachusetts, Amherst, MA 01003, USA, edwardc@schoolph.umass.edu

Tae Gen Son Laboratory of Neurosciences, National Institutes of Health, National Institute on Aging, Baltimore, MD 21224, USA, sont2@grc.nia.nih.gov

Roy G. Cutler Laboratory of Neurosciences, National Institutes of Health, National Institute on Aging, Baltimore, MD 21224, cutlerro@grc.nia.nih.gov

Simonetta Camandola Laboratory of Neurosciences, National Institute on Aging, Intramural Research Program, Baltimore, MD 21224, camandolasi@grc.nia.nih.gov

Wayne Chadwick Receptor Pharmacology Unit, National Institute on Aging, National Institutes of Health, Biomedical Research Center, Baltimore, MD 21224, USA, chadwickwa@nia.nih.gov

Stuart Maudsley Laboratory of Neurosciences, National Institutes of Health, National Institute on Aging, Baltimore, MD 21224, USA, maudsleyst@grc.nia.nih.gov

Alexis M. Stranahan Department of Psychological and Brain Sciences, Johns Hopkins University, Baltimore, MD 21218, USA, alexis.stranahan@jhu.edu

Bronwen Martin Laboratory of Clinical Investigation, National Institutes of Health, National Institute on Aging, Baltimore, MD 21224, USA, martinbro@mail.nih.gov

Sunggoan Ji Laboratory of Neurosciences, National Institutes of Health, National Institute on Aging, Baltimore, MD 21224, USA, jis2@mail.nih.gov

Caitlin M. White Laboratory of Neurosciences, National Institutes of Health, National Institute on Aging, Baltimore, MD 21224, USA, whitecm2@mail.nih.gov

Suresh I. S. Rattan Laboratory of Cellular Aging, Department of Molecular Biology, University of Aarhus, DK 8000 Aarhus C, Denmark, rattan@mb.au.dk

Dino Demirovic Laboratory of Cellular Aging, Department of Molecular Biology, University of Aarhus, DK8000 Aarhus C, Denmark, rattan@mb.au.dk

Hormesis: What It Is and Why It Matters

Mark P. Mattson and Edward J. Calabrese

Abstract Hormesis describes any process in which a cell, organism, or group of organisms exhibits a biphasic response to exposure to increasing amounts of a substance or condition (e.g., chemical, sensory stimulus, or metabolic stress); typically, low-dose exposures elicit a stimulatory or beneficial response, whereas high doses cause inhibition or toxicity. The biphasic dose-response signature of hormesis is a common result of experiments in the field of toxicology, but the low-dose data have been largely ignored, and the prevailing view is that it is important to reduce levels of toxins as much as possible. However, in many cases, the "toxins" actually have essential or beneficial effects in low amounts. Prominent examples of such beneficial "toxins" are trace metals such as selenium, chromium, and zinc. Fundamental inter- and intracellular signals also exhibit hormetic dose responses, including hormones, neurotransmitters, growth factors, calcium, and protein kinases. Moreover, everyday health-promoting lifestyle factors, including exercise and dietary energy restriction, act, at least in part, through hormetic mechanisms involving activation of adaptive cellular stress response pathways (ACSRPs). ACSRPs typically involve receptors coupled to kinases and activation of transcription factors that induce the expression of cytoprotective proteins such as antioxidant enzymes, protein chaperones, and growth factors. The recognition and experimental utilization of hormesis is leading to novel approaches for preventing and treating a range of diseases, including cancers, cardiovascular disease, and neurodegenerative disorders.

Keywords Adaptation · Biphasic · Environmental protection · Evolution · Preconditioning · Stress · Toxins

---

M.P. Mattson (✉)
Laboratory of Neurosciences, National Institute on Aging, Intramural Research Program, Baltimore, MD 21224, USA
e-mail: mattsonm@grc.nia.nih.gov

M.P. Mattson, E.J. Calabrese, *Hormesis*, DOI 10.1007/978-1-60761-495-1_1,

Hormesis Is a Fundamental Feature of Biological Systems

A defining characteristic of hormesis is a biphasic dose-response curve, with beneficial or stimulatory effects at low doses and adverse or inhibitory effects at high doses. Biphasic responses to increasing doses of chemicals have been widely reported for a range of agents (mercury, arsenic, pesticides, radiation, etc.) and organisms (bacteria, worms, flies, rodent, humans, and many others). In fact, toxins more often exhibit a hormetic dose response (low-dose stimulation or beneficial effect, and high-dose inhibition or toxicity) than they do a linear dose response (toxicity proportional to the level of exposure). Calabrese has cataloged thousands of examples of hormetic dose responses in the fields of biology, toxicology, and medicine (Calabrese and Blain, 2005; Cook and Calabrese, 2006; and see the chapter in this book, Hormesis: Once Marginalized, Evidence Now Supports Hormesis as the Most Fundamental Dose Response). Examples of hormetic dose response include the following: low amounts of cadmium improve the reproductive capacity of snails, whereas high doses are lethal (Lefcort et al., 2008); low doses of radiation increase the growth rate of plants and can increase the lifespan of mice (Luckey, 1999); and chemicals that can cause cancer when consumed in high amounts can actually inhibit cancer cell growth when taken in low doses (Calabrese, 2005).

Paracelsus recognized four centuries ago that drugs are actually toxins that have beneficial effects at low doses (Fig. 1). The biphasic dose-response relation is not limited to exposures to environmental agents and drugs, however; it permeates biology, physiology, and the daily experiences of all organisms. Well-known categories of agents that exert biphasic effects on human health are minerals and vitamins. Selenium, a trace element obtained in the diet, is essential for health because it is necessary for the proper function of at least 30 selenoproteins (Dodig and Cepelak, 2004). However, high levels of selenium are toxic and can even cause death. Vitamin D is critical for the growth and health of bones and for wound healing, among other processes, but excessive intake of vitamin D can cause hypercalcemia and associated pathologies in the kidneys and other organs (Vieth, 2007). Vitamin A is necessary for proper development of multiple organs and for maintenance of the health of the eye and other tissues in the adult; however, excessive intake of vitamin A can cause liver damage, may promote osteoporosis, and may also adversely affect the cardiovascular system (Penniston and Tunumihardjo, 2006). Iron is essential for red blood cell health and also serves important regulatory functions in other cell types, but excessive iron intake can cause oxidative damage to tissues (Van Gossum and Neve, 1998).

Another example of hormesis centers on glutamate, an amino acid neurotransmitter that is critical for the transfer of electrical activity from one nerve cell to another in the brain. The relatively low amounts of glutamate released at the synapse when the brain is engaged in activities such as reading and writing activate adaptive cellular stress response pathways (ACSRPs) that benefit the nerve cells, promoting their growth and survival (Fig. 2). However, excessive amounts of glutamate can damage and kill nerve cells in a process called excitotoxicity that occurs during

"All things are poison and nothing is without poison, only the dose permits something not to be poisonous" -Paracelsus

Fig. 1 Paracelsus was a Swiss-born alchemist and physician who pioneered the use of chemicals and minerals in medicine. He recognized the importance of the dose of chemicals in determining whether they are therapeutic or toxic, and essentially predicted the prevalence of the biphasic nature of the dose-response curve as typical of all medicines

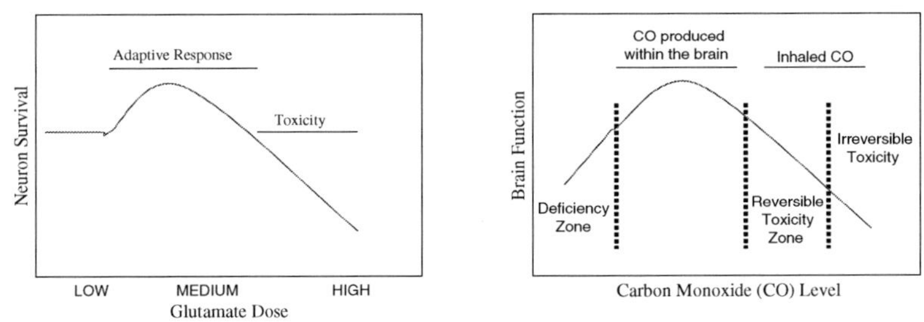

Fig. 2 Hormetic dose responses of nerve cells to the neurotransmitter glutamate and the gaseous messenger carbon monoxide (CO). Low to medium doses of glutamate mediate synaptic transmission and plasticity, learning and memory, and other behaviors. High amounts of glutamate can cause excessive calcium influx into neurons, resulting in neuronal damage and death; this occurs in epilepsy and stroke and may also occur in Alzheimer's, Parkinson's, and Huntington's diseases. Carbon monoxide is produced by cells in the brain and plays important roles in signaling within and between neurons and in blood vessel cells. Inhaled CO can result in levels in blood and tissues that, if sustained, can cause asphyxia and death

severe epileptic seizures, as well as in Alzheimer's and Parkinson's diseases. Carbon monoxide also exerts hormetic effects on cells and organisms. Carbon monoxide is widely known as a toxic gas present in the exhaust of combustion engines, but carbon monoxide also is produced by cells in the body, where it serves important signaling functions promoting blood vessel relaxation and communication between nerve cells (Kaczorowski and Zuckerbraun, 2007; Fig. 2).

Another class of hormetic molecules in cells comprises oxygen free radicals, notorious for their ability to damage DNA, proteins, and membrane lipids. Free radicals are believed to play major roles in the aging process and in various diseases, including cardiovascular and inflammatory diseases, cancers, and neurodegenerative disorders (Giacosa and Filiberti, 1996; Mattson and Liu, 2002). Recent research has clearly shown, however, that low amounts of some free radicals serve important functions in cells that involve the activation of ACSRPs (Ridnour et al., 2006; Valko et al., 2007. One example is superoxide anion radical ($O_2^{-\cdot}$), which is produced by the activity of the mitochondrial electron transport chain as a byproduct of oxidative phosphorylation (the process that produces adenosine triphosphate [ATP], the major cellular energy substrate). Superoxide is normally "detoxified" by the actions of superoxide dismutases, which convert $O_2^{-\cdot}$ to hydrogen peroxide; hydrogen peroxide is then converted to water by the actions of catalase and glutathione peroxidase. Thus, levels of $O_2^{-\cdot}$ are normally kept low. However, high amounts of $O_2^{-\cdot}$ can occur in certain conditions (e.g., with reductions in levels of antioxidant enzymes) and can damage cells by conversion to more highly reactive free radicals, including hydroxyl radical and (by interaction with nitric oxide) peroxynitrite (Mattson, 2004). In response to physiological signals such as neurotransmitters, cytokines, and calcium fluxes, $O_2^{-\cdot}$ is produced and mediates the activation of kinases and transcription factors (Camello-Almaraz et al., 2006; Kishida and Klann, 2007). Reactive oxygen species such as $O_2^{-\cdot}$ also mediate responses of immune cells. For example, in response to exogenous (allergens) and endogenous (molecules released from damaged cells) factors, mast cells generate $O_2^{-\cdot}$ and other free radicals that induce degranulation, leukotriene secretion, and cytokine production (Suzuki et al., 2005). Free radicals also play important roles in signaling processes that regulate vascular endothelial cell function and blood pressure (Wolin, 1996). Mitochondria normally produce $O_2^{-\cdot}$ in bursts or "flashes" (Wang et al., 2008), and one possible function of such $O_2^{-\cdot}$ flashes is to activate adaptive cellular stress response signaling pathways.

Hormesis Is a Manifestation of a Fundamental Feature of Evolution

To survive and propagate, organisms must be able to withstand various hazards in their environment and outcompete their rivals for limited energy resources. Mechanisms for responding adaptively to stress are fundamental to the process of evolution and are therefore encoded in the genomes of all organisms. Early life forms lived in hostile environments where they were subjected to a range of toxic metals, ultraviolet light, and large changes in temperature. Survival was favored in organisms that were able to resist these environmental stressors. Various examples of adaptations to stress selected for during evolution are presented in the chapter in this book, The Fundamental Role of Hormesis in Evolution. One fundamental means of coping with exposures to potentially lethal environmental conditions is to move away from the hazard, which is presumably one driving force for the evolution of cell and organismal motility. Alternatively, the ability to change physiological processes to withstand the noxious agent would have allowed the organism to remain in its location. Moreover, by responding adaptively to low levels of various environmental stressors, organisms were able to expand the range of environments in which they could survive. For example, levels of arsenic in soils and drinking water vary considerably across the globe, and in some areas levels are high enough to cause sickness and death (Mukherjee et al., 2006). However, low doses of arsenic can protect cells against oxidative stress and DNA damage (Snow et al., 2005), indicating the existence of a biphasic (hormetic) profile of arsenic exposure in which low doses may activate an adaptive stress response that can protect against stress and disease.

Cells and organisms that were vulnerable to specific environmental factors evolved to become resistant to the factors. Moreover, in many instances, the organisms evolved in ways that allowed them to utilize "toxic" elements and molecules to their advantage. One excellent example mentioned earlier is selenium, which is toxic at high doses and, early in evolution, was likely toxic at lower doses. During evolution, selenium began to be used by organisms to enhance the function of certain enzymes, and selenium is now required for the health and survival of many organisms, including humans (Boosalis, 2008). The calcium ion (Ca^{2+}) is widely known for its fundamental role in intracellular signaling and as a mediator of a wide range of cell responses, including proliferation, differentiation, motility, and secretion (Schreiber, 2005). However, the excessive accumulation of Ca^{2+} in cells can cause dysfunction and death of the cells, a process implicated in many diseases, including neurodegenerative disorders and cardiovascular disease (Allen et al., 1993; Mattson, 2007). Thus, cells have evolved a battery of mechanisms to guard against excessive Ca^{2+} accumulation, including Ca^{2+} channels and Ca^{2+} pumps in the plasma and endoplasmic reticulum membranes and Ca^{2+}-binding proteins (Fig. 3). Complex arrays of Ca^{2+}-regulating mechanisms and Ca^{2+}-mediated signaling pathways have evolved to serve the most sophisticated functions of higher organisms, including the events that occur at synapses (neurotransmitter release and postsynaptic responses to neurotransmitters) that are the basis of cognition, reasoning, and the planning of survival strategies (Blitzer et al., 2005). There are may other examples of potentially toxic chemicals that serve critical physiological functions in low concentrations or controlled (transient) higher doses as occur during Ca^{2+} influx and removal in excitable cells.

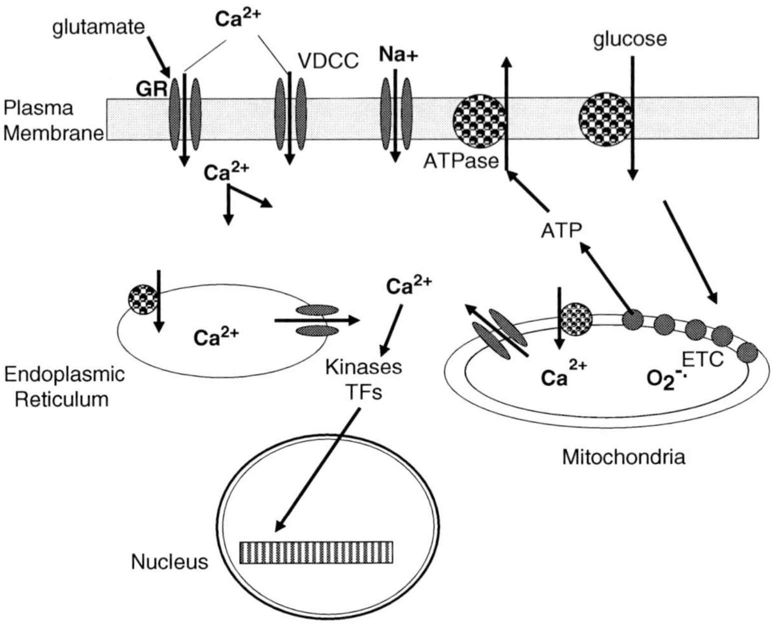

Fig. 3 Calcium signaling pathways and systems that regulate Ca^{2+} levels and movements in cells. The concentration of Ca^{2+} is much higher outside the cell (1–2 mM) than in the cytoplasm (typically 100–300 nM). This gradient is established by a plasma membrane that is impermeable to Ca^{2+} but contains ATP-dependent pumps (Ca^{2+}-ATPase) that extrude Ca^{2+}. The plasma membrane also contains voltage-dependent Ca^{2+} channels (VDCC) and ligand-gated Ca^{2+} channels such as the *N*-methyl-D-aspartate type of glutamate receptor (GR). The Ca^{2+} that enters cells through the latter channels functions as a signal that regulates a range of cellular responses, including proliferation, differentiation, motility, and gene expression through the activation of kinases and transcription factors (TFs). Ca^{2+} is transported into the endoplasmic reticulum via the activity of the sarco/endoplasmic reticulum Ca^{2+}-ATPase (SERCA)

Cellular and Molecular Mediators of Hormetic Responses

How can exposures to low levels of a toxin or other stressful agent result in beneficial effects on cells? Many different signaling pathways have been shown to mediate adaptive stress responses in cells and organisms, and there are undoubtedly many more that remain to be discovered. Typically these hormetic pathways involve sensor molecules, intracellular messengers, and transcription factors that induce the expression of genes that encode cytoprotective proteins. The importance of such stress resistance proteins in evolution is exemplified by the fact that a large portion of the genes in the genome are involved in stress responses (Cooper et al., 2003). Several different ACSRPs that mediate hormetic responses to oxidative stress have been described, including the Nrf-2–ARE pathway (Kang et al., 2005) and the sirtuin–FOXO pathway (Jiang, 2008). These pathways each culminate in the nucleus, where they induce the expression of genes encoding an array of proteins that protect cells against stress, including antioxidant enzymes, protein chaperones, and proteins involved in energy metabolism. These pathways may be activated rather directly by chemicals. For example, sulforaphane, a chemical present in high amounts in broccoli, can interact with Nrf-2; resveratrol (present in red grapes and wine) activates the sirtuin pathway; and allicin (a chemical in garlic and onions) can activate membrane TRP channels, resulting in calcium influx (Mattson and Cheng, 2006).

On the other hand, many potentially toxic substances and conditions activate ACSRPs indirectly by inducing a nonspecific oxidative, metabolic, or ionic stress. Two transcription factors that are activated in many cell types in response to oxidative and metabolic stress are NF-κB (Mattson and Meffert, 2006) and hypoxia-inducible factor 1 (HIF1; Loor and Schumacker, 2008). NF-κB coordinates cellular responses to infection and tissue injury throughout the body. Activation of NF-κB in immune cells such as lymphocytes and macrophages induces the production of cytokines such as tumor necrosis factor that function in destroying infectious agents and removing dead cells in injured tissues. Activation of NF-κB in cells such as neurons promotes their survival by inducing the expression of manganese superoxide dismutase and Bcl-2, for example (Mattson and Meffert, 2006). Whereas low levels of NF-κB activation are beneficial, high sustained activation can cause pathological damage to tissues. HIF1 responds to hypoxia and increased cellular energy demand as occurs in muscle cells during exercise (Freyssenet, 2007).

One class of highly specialized molecular bodyguards that mediate hormetic responses is made up of the heat-shock proteins, which serve as chaperones that protect other proteins against damage (Kim et al., 2006). The production of heat-shock proteins is rapidly increased not only by high temperatures, but also under conditions of oxidative and metabolic stress as occur during exposures to chemical toxins or tissue inflammation. The heat-shock proteins then bind to vulnerable proteins in different parts of the cell and shield them from attack by oxygen free radicals and other damaging chemicals. Some molecular bodyguards function as messengers that leave the neuron exposed to the threat and alert adjacent neurons of the danger. Growth factors are one such early warning system—they mobilize the defenses of cells that are within the war zone but not yet under attack. For example, in response to multiple stressors, including exercise, ischemia, and exposure to certain "excitotoxins," brain cells produce several different growth factors that promote the survival of their neighbors, including fibroblast growth factor, nerve growth factor, and brain-derived neurotrophic factor (Mattson et al., 1995).

Hormesis in Medicine: Dose and Frequency of Treatment Are Both Important

Most, if not all, drugs exhibit hormetic dose responses, with beneficial effects at therapeutic doses and toxic effects with overdoses. The toxic effects of high doses may be due to a higher level of action (inhibition or stimulation) at the specific molecular target of the drug (typically a receptor or enzyme) or may result from nonspecific effects on metabolism. For example, low doses of β-adrenergic receptor antagonists are effective in reducing blood pressure, whereas higher concentrations can cause circulatory collapse (Love and Elshami, 2002). At therapeutic doses, γ–aminobutyric acid (GABA) receptor agonists such as diazepam (Valium) are effective in reducing anxiety, whereas at higher concentrations they adversely affect cognition and motor function (Gorenstein et al., 1994). Aspirin at low doses is effective in preventing myocardial infarction by inhibiting platelet aggregation and clot formation; higher doses can reduce pain by inhibiting prostaglandin production but have the adverse effect of promoting ulcer formation (Vane and Botting, 2003).

Some commonly prescribed drugs may exert their beneficial actions by hormetic mechanisms. One example comes from studies of psychiatric disorders. Antidepressants such as fluoxetine (Prozac) and paroxetine (Paxil) stimulate nerve cells to produce brain-derived neurotrophic factor (BDNF), a protein that promotes the growth and survival of neurons. Patients who do not respond to antidepressants may benefit from a more dramatic hormetic treatment called electroconvulsive shock therapy in which nerve cells are vigorously stimulated by passing an electric current through the brain. The widely prescribed diabetes drug metformin may act, in part, by inducing a mild stress in the muscle cells similar to what occurs during exercise. Both exercise and metformin stimulate the activity of a protein called AMP-activated protein kinase (AMPK), resulting in increased sensitivity of muscle cells to insulin.

Not only is the dose a critical determinant of whether an environmental challenge is beneficial or damaging, but in addition the frequency of exposure is key because cells must have time to recover to benefit from the stress. The importance of a recovery period for the accrual of the benefits of exercise is widely recognized. Less well known is the importance of a recovery period for the beneficial effects of many other hormetic stressors, including dietary energy restriction, phytochemicals, and even certain drugs. Although fasting has been part of many religions for thousands of years, its far-reaching health benefits were brought to public attention with the publication of Upton Sinclair's *The Fasting Cure* in 1911. Sinclair described his experiences and those of several hundred other people whose various maladies were "cured" by fasting. Studies have demonstrated the ability of regular fasting to improve the health and function of major organs, including the brain and heart (Bruce-Keller et al., 1999; Duan et al., 2003; Maswood et al., 2004; Wan et al., 2003a, 2003b; Mager et al., 2006). The mild stress that occurs during fasting is important for its beneficial effects, as is a refeeding recovery period to provide the nutrients necessary for maintaining tissue and organ functions. A major goal of the fields of pharmacology and medicine should therefore be to establish the dose and frequency of drug administration that maximize relief of symptoms while minimizing side effects. Unfortunately, the most common approach, that has also been applied to dietary supplements, assumes that a chemical is most effective when its concentration in the body is maintained constant. However, this notion may not apply to drugs and dietary components or supplements that act by a hormetic mechanism. Instead, many chemicals may provide an optimal therapeutic benefit when delivered in a pulsatile or intermittent manner that allows a recovery period for cells to respond adaptively to the stress induced by the chemical.

Are Beneficial Chemicals in Fruits and Vegetables Toxins Acting at Low Doses?

Emerging evidence suggests that some drugs and health-promoting chemicals in fruits and vegetables may exert their beneficial effects by activating ACSRPs. The evidence that consumption of fruits and vegetables is associated with a reduced risk for cardiovascular disease, certain cancers, and some neurodegenerative disorders has resulted in efforts to identify the specific chemicals responsible for these health benefits. Because damage caused by free radicals is involved in most major diseases, it has been widely believed that the direct antioxidant activity of phytochemicals is responsible for their beneficial effects. However, most phytochemicals are only effective as antioxidants when they are present in very high concentrations that are not achievable by eating normal amounts of fruits and vegetables, and there is often a biphasic dose-response relationship for phytochemicals (low-dose beneficial effects and high-dose toxic effects), which argues against an antioxidant mechanism of action. Moreover, several major clinical trials failed to demonstrate beneficial effects of high doses of antioxidants for the treatment of cancers, cardiovascular disease, and Alzheimer's disease. Based on this kind of information, evolutionary considerations, and our research, we believe that instead of a direct antioxidant mechanism, many phytochemicals exert their health benefits by inducing mild stress responses in cells.

One important evolutionary adaptation of plants is the ability to produce toxic substances and concentrate them in regions such as the skin of fruits and the buds of leaves to dissuade insects and other organisms from eating them. Hundreds of these "natural biopesticides" exist but are insufficient in the amounts normally consumed in our diets to achieve toxic concentrations in the body. Instead, the phytochemicals activate one or more specific adaptive stress response signal transduction pathways and transcription factors (Mattson and Cheng, 2006). For example, chemicals present in broccoli (sulforaphane) and curry spice (curcumin) activate a protein located in the cytoplasm called Nrf-2, which then moves to the nucleus, where it activates genes for antioxidant enzymes and so-called "phase 2 detoxification" enzymes. A different hormetic pathway was recently found to be activated by resveratrol, a phytochemical believed to be responsible for the health benefits of red grapes and wine. Resveratrol activates sirtuin-1, which in turn stimulates a transcription factor called FOXO, resulting in the production of proteins that counteract oxidative stress. Other phytochemicals, including allicin (in garlic) and capsaicin (in hot peppers), induce a mild stress response in cells by causing the opening of pores in the cell membrane called transient receptor potential (TRP) channels, resulting in the influx of calcium. The calcium then activates a transcription factor called the cAMP-response element–binding protein (CREB), which induces the production of

BDNF and other growth factors. Activation of these different pathways by phytochemicals can protect cells against stress and thereby help them to avoid injury and disease.

Hormesis Is Not Homeopathy

Homeopathy is a 200-year-old theory of medicine based on the work of Samuel Hahnemann that proposes that agents that produce symptoms of a disease in a healthy person could be used to treat ill patients. From this is derived the well-known principle of homeopathy that "like cures like." Hahnemann believed that his treatments could be effective at vanishingly low doses, a possibility that generated skepticism within his homeopathic medical community, as well as within the broader biomedical community. Homeopathy and the concept of hormesis became linked through the work of Hugo Schulz at the University of Greiswald in northern Germany. In the mid 1880s Schulz observed that chemical disinfectants stimulated the metabolism of yeast at low doses while being inhibitory at higher doses. Schulz immediately thought that he had discovered the scientific principle underlying the medical practice of homeopathy. He advocated this perspective until his death in 1932. In general, the work of Schulz had no connection with homeopathy. It was based on assessing the dose-response continuum, that is, doses that exceeded the toxic threshold and doses immediately below it. The hormetic dose response is a normal component of the traditional dose response. Large amounts of experimentally derived data have demonstrated that adaptive responses are observable at doses immediately below toxic thresholds. This is the hormetic zone, not a dose zone multiple orders of magnitude below the threshold and into a vanishingly low concentration at which molecules may or may not even be present. Thus, the biological process of hormesis is only linked to the purely human construct of homeopathy because of a mistake by Hugo Schulz.

Implications of Hormesis for the Practices of Environmental Protection and Medicine

Ignorance is not bliss. As described and documented throughout the chapters of this book, the prevalence of hormesis in biological systems demands that data from full dose-response studies be available to inform those who make decisions regarding the management of environmental hazards and the treatment of patients. Many chemicals in the environment, particularly those that are natural, although toxic at high doses, exert beneficial effects at low doses. Examples include metals (selenium, zinc, iron, etc.), phytochemicals (quercetin, curcumin, sulforaphane, etc.) and gases (oxygen, carbon monoxide, ozone, etc.). The goal should therefore be to establish the hormetic range of doses and then take measures to constrain exposures to doses within this optimal range. Eliminating a toxic chemical from the environment without knowing about its biological effects at low doses may result in poorer health outcomes compared to reducing levels of the chemical to within the hormetic dose range.

In drug development the usual approach for deciding on a dose of medicine is to first determine the minimum dose at which toxicity is observed and then set the therapeutic dose somewhat below the toxic dose. In many cases, the resulting "therapeutic" dose may actually coincide with the hormetic dose. For example, therapeutic doses of antidepressants such as fluoxetine and paroxetine induce an adaptive stress response in neurons in the brain that results in stimulation of the expression of BDNF (Martinowich and Lu, 2008). BDNF promotes the growth, plasticity, and survival of neurons and also induces the production of new neurons from stem cells in the hippocampus (Mattson et al., 2004). Of interest, the antidepressant effect of moderate exercise may be mediated by a similar hormetic mechanism involving BDNF (Li et al., 2008). However, in other cases, the treatment dose may not be the most effective dose, particularly in cases in which the drug acts by a hormetic mechanism. For example, very low doses of aspirin (well below doses that are toxic) reduce the risk of myocardial infarction and stroke (Webster and Douglas, 1987; Hennekens, 2002). It will be of considerable interest, and of potential clinical importance, to reevaluate many commonly used drugs in the low (possibly hormetic) dose range. Incorporation of low doses studied in preclinical models may also identify agents with "off-target" low-dose beneficial actions.

Acknowledgments This work was supported by the Intramural Research Program of the National Institute on Aging, National Institutes of Health.

References

Allen DG, Cairns SP, Turvey SE, Lee JA (1993) Intracellular calcium and myocardial function during ischemia. Adv Exp Med Biol 346: 19–29.

Blitzer RD, Iyengar R, Landau EM (2005) Postsynaptic signaling networks: cellular cogwheels underlying long-term plasticity. Biol Psychiatry 57: 113–119.

Boosalis MG (2008) The role of selenium in chronic disease. Nutr Clin Pract 23: 152–160.

Bruce-Keller AJ, Umberger G, McFall R, Mattson MP (1999) Food restriction reduces brain damage and improves behavioral outcome following excitotoxic and metabolic insults. Ann Neurol 45: 8–15.

Calabrese EJ (2005) Cancer biology and hormesis: human tumor cell lines commonly display hormetic (biphasic) dose responses. Crit Rev Toxicol 35: 463–582.

Calabrese EJ, Blain R (2005) The occurrence of hormetic dose responses in the toxicological literature, the hormesis database: an overview. Toxicol Appl Pharmacol 202: 289–301.

Camello-Almaraz C, Gomez-Pinilla PJ, Pozo MJ, Camello PJ (2006) Mitochondrial reactive oxygen species and Ca2+ signaling. Am J Physiol Cell Physiol 291: C1082–C1088.

Cook R, Calabrese EJ (2006) The importance of hormesis to public health. Environ Health Perspect 114: 1631–1635.

Cooper B, Clarke JD, Budworth P, Kreps J, Hutchison D, Park S, Guimil S, Dunn M, Luginbühl P, Ellero C, Goff SA, Glazebrook J (2003) A network of rice genes associated with stress response and seed development. Proc Natl Acad U S A. 100: 4945–4950.

Dodig S, Cepelak I (2004) The facts and controversies about selenium. Acta Pharm 54: 261–276.

Duan W, Guo Z, Jiang H, Ware M, Li XJ, Mattson MP (2003) Dietary restriction normalizes glucose metabolism and BDNF levels, slows disease progression, and increases survival in huntingtin mutant mice. Proc Natl Acad Sci USA 100: 2911–2916.

Freyssenet D (2007) Energy sensing and regulation of gene expression in skeletal muscle. J Appl Physiol 102: 529–540.

Giacosa A, Filiberti R (1996) Free radicals, oxidative damage and degenerative diseases. Eur J Cancer Prev 5: 307–312.

Gorenstein C, Bernik MA, Pompéia S (1994) Differential acute psychomotor and cognitive effects of diazepam on long-term benzodiazepine users. Int Clin Psychopharmacol 9: 145–153.

Hennekens CH (2002) Update on aspirin in the treatment and prevention of cardiovascular disease. Am J Manag Care 8: S691–S700.

Jiang WJ (2008) Sirtuins: novel targets for metabolic disease in drug development. Biochem Biophys Res Commun 373: 341–344.

Kaczorowski DJ, Zuckerbraun BS (2007) Carbon monoxide: medicinal chemistry and biological effects. Curr Med Chem 14: 2720–2725.

Kang KW, Lee SJ, Kim SG (2005) Molecular mechanism of nrf2 activation by oxidative stress. Antioxid Redox Signal 7: 1664–1673.

Kim HP, Morse D, Choi AM (2006) Heat-shock proteins: new keys to the development of cytoprotective therapies. Expert Opin Ther Targets 10: 759–769.

Kishida KT, Klann E (2007) Sources and targets of reactive oxygen species in synaptic plasticity and memory. Antioxid Redox Signal 9: 233–244.

Lefcort H, Freedman Z, House S, Pendleton M (2008) Hormetic effects of heavy metals in aquatic snails: is a little bit of pollution good?. Ecohealth 5: 10–17.

Li Y, Luikart BW, Birnbaum S, Chen J, Kwon CH, Kernie SG, Bassel-Duby R, Parada LF (2008) TrkB regulates hippocampal neurogenesis and governs sensitivity to antidepressive treatment. Neuron 59: 399–412.

Loor G, Schumacker PT (2008) Role of hypoxia-inducible factor in cell survival during myocardial ischemia-reperfusion. Cell Death Differ 15: 6866–6890.

Love JN, Elshami J (2002) Cardiovascular depression resulting from atenolol intoxication. Eur J Emerg Med 9: 111–114.

Luckey TD (1999) Nurture with ionizing radiation: a provocative hypothesis. Nutr Cancer 34: 1–11.

Mager DE, Wan R, Brown M, Cheng A, Wareski P, Abernethy DR, Mattson MP (2006) Caloric restriction and intermittent fasting alter spectral measures of heart rate and blood pressure variability in rats. FASEB J 20: 631–637.

Martinowich K, Lu B (2008) Interaction between BDNF and serotonin: role in mood disorders. Neuropsychopharmacology 33: 73–83.

Maswood N, Young J, Tilmont E, Zhang Z, Gash DM, Gerhardt GA, Grondin R, Roth GS, Mattison J, Lane MA, Carson RE, Cohen RM, Mouton PR, Quigley C, Mattson MP, Ingram DK (2004) Caloric restriction increases neurotrophic factor levels and attenuates neurochemical and behavioral deficits in a primate model of Parkinson's disease. Proc Natl Acad Sci USA 101: 18171–18176.

Mattson MP, Lovell MA, Furukawa K, Markesbery WR (1995) Neurotrophic factors attenuate glutamate-induced accumulation of peroxides, elevation of intracellular Ca2+ concentration, and neurotoxicity and increase antioxidant enzyme activities in hippocampal neurons. J Neurochem 65: 1740–1751.

Mattson MP, Liu D (2002) Energetics and oxidative stress in synaptic plasticity and neurodegenerative disorders. Neuromolecular Med 2: 215–231.

Mattson MP (2004) Metal-catalyzed disruption of membrane protein and lipid signaling in the pathogenesis of neurodegenerative disorders. Ann N Y Acad Sci 1012: 37–50.

Mattson MP, Maudsley S, Martin B (2004) BDNF and 5-HT: a dynamic duo in age-related neuronal plasticity and neurodegenerative disorders. Trends Neurosci 27: 589–594.

Mattson MP, Cheng A (2006) Neurohormetic phytochemicals: low-dose toxins that induce adaptive neuronal stress responses. Trends Neurosci 29: 632–639.

Mattson MP, Meffert MK (2006) Roles for NF-kappaB in nerve cell survival, plasticity, and disease. Cell Death Differ 13: 852–860.

Mattson MP (2007) Calcium and neurodegeneration. Aging Cell 6: 337–350.

Mukherjee A, Sengupta MK, Hossain MA, Ahamed S, Das B, Nayak B, Lodh D, Rahman MM, Chakraborti D (2006) Arsenic contamination in groundwater: a global perspective with emphasis on the Asian scenario. J Health Popul Nutr 24: 142–163.

Penniston KL, Tanumihardjo SA (2006) The acute and chronic toxic effects of vitamin A. Am J Clin Nutr 83: 191–201.

Ridnour LA, Thomas DD, Donzelli S, Espey MG, Roberts DD, Wink DA, Isenberg JS (2006) The biphasic nature of nitric oxide responses in tumor biology. Antioxid Redox Signal 8: 1329–1337.

Schreiber R (2005) Ca2+ signaling, intracellular pH and cell volume in cell proliferation. J Membr Biol 205: 129–137.

Snow ET, Sykora P, Durham TR, Klein CB (2005) Arsenic, mode of action at biologically plausible low doses: what are the implications for low dose cancer risk?. Toxicol Appl Pharmacol 207: 557–564.

Suzuki Y, Yoshimaru T, Inoue T, Niide O, Ra C (2005) Role of oxidants in mast cell activation. Chem Immunol Allergy 87: 32–42.

Valko M, Leibfritz D, Moncol J, Cronin MT, Mazur M, Telser J (2007) Free radicals and antioxidants in normal physiological functions and human disease. Int J Biochem Cell Biol 39: 44–84.

Vane JR, Botting RM (2003) The mechanism of action of aspirin. Thromb Res 110: 255–258.

Van Gossum A, Neve J (1998) Trace element deficiency and toxicity. Curr Opin Clin Nutr Metab Care 1: 499–507.

Vieth R (2007) Vitamin D toxicity, policy, and science. J Bone Miner Res 22(Suppl 2): V64–V68.

Wan R, Camandola S, Mattson MP (2003a) Intermittent fasting and dietary supplementation with 2-deoxy-D-glucose improve functional and metabolic cardiovascular risk factors in rats. FASEB J 17: 1133–1134.

Wan R, Camandola S, Mattson MP (2003b) Intermittent food deprivation improves cardiovascular and neuroendocrine responses to stress in rats. J Nutr 133: 1921–1929.

Wang W, Fang H, Groom L, Cheng A, Zhang W, Liu J, Wang X, Li K, Han P, Zheng M, Yin J, Wang W, Mattson MP, Kao JP, Lakatta EG, Sheu SS, Ouyang K, Chen J, Dirksen RT, Cheng H (2008) Superoxide flashes in single mitochondria. Cell 134: 279–290.

Webster J, Douglas AS (1987) Aspirin and other antiplatelet drugs in the prophylaxis of thrombosis. Blood Rev 1: 9–20.

Wolin MS (1996) Reactive oxygen species and vascular signal transduction mechanisms. Microcirculation 3: 1–17.

Hormesis: Once Marginalized, Evidence Now Supports Hormesis as the Most Fundamental Dose Response

Edward J. Calabrese

Abstract The biomedical community made a fundamental error on the nature of the dose-response relationship early in the 20th century and has perpetuated this error to the present. The error was the byproduct of the conflict between homeopathy and traditional medicine. To deny support to homeopathy, leaders of the biomedical community rejected the hormetic biphasic dose-response model, the proposed explanatory principle of homeopathy. The threshold dose-response model was adopted as an alternative model, quickly becoming central to toxicology/pharmacology and their numerous applications. Despite its near-universal acceptance, no attempt was made to validate the ability of the threshold model to accurately predict responses in the below-threshold zone at the time of acceptance and throughout the 20th century. In contrast, the hormetic biphasic dose-response model became marginalized and was excluded from the mainstream of pharmacological/toxicological teaching and practice, textbook development, professional society journal publications, annual meeting presentations, grant funding, and use in government risk assessment. Over the last decade there has been a resurgence of interest in hormesis due to findings indicating that hormetic responses are common, reproducible, and generalizable, as well as independent of biological model, endpoint, and chemical class/physical stressor. Large-scale studies have indicated that the threshold model fails to accurately predict responses below the threshold, whereas the hormetic dose-response model performs very well. These findings indicate that the biomedical community made an error on the nature of the dose-response relationship, compromising the accuracy of toxicological and risk assessment practices, including environmental exposure standards, and impeding drug discovery/development and drug safety studies.

Keywords Hormesis · Hormetic · Biphasic · U-shaped · J-shaped · Dose-response relationship · Adaptive response · Preconditioning · History of science

E.J. Calabrese (✉)
Department of Environmental Health Sciences, School of Public Health and Health Sciences, University of Massachusetts, Amherst, MA 01003, USA
e-mail: edwardc@schoolph.umass.edu

M.P. Mattson, E.J. Calabrese, *Hormesis*, DOI 10.1007/978-1-60761-495-1_2,

Introduction

The dose-response relationship is the central concept within the fields of pharmacology and toxicology. It guides how studies are designed and biostatistical modeling is performed, the general focus of mechanistic research, drug efficacy and safety evaluation, and governmental environmental risk assessment practices for protecting humans and other life against threats to food, water, air, and soil. The dose-response relationship is also a fundamental concept of biology, in that it is central to evolutionary theory and its underlying processes of mutation, DNA repair, and a plethora of integrative adaptive responses. Central to the biological and health sciences, the dose-response relationship is a scientific concept that seems as obvious as it is profound, being nearly universally understood based on common experience. Therein lies the trap into which the general public and the scientific community have fallen.

Over the last century the scientific community accepted the threshold dose-response model as a description of how chemical and physical stressor agents affect the vast range of biological processes across essentially all forms of life and biological organization. This concept has become integrated into all biological disciplines and regulatory practices, quietly evolving into a fundamental concept. Reinforcing this "scientific" decision on the primacy of the threshold dose-response model is the general recognition of thresholds in the physical sciences, such as melting, boiling, and freezing points, and common experiences with medications and other products. The convergence of agreement on the dose-response model by the scientific community and the general public is also as important as it is in reinforcing belief in the validity of this concept, hence its acceptance and status as a central pillar in various disciplines.

Despite the history of science with its self-correcting features and the wisdom of the general public's experiences and its integration of perceptions concerning the dose-response relationship, both science and the lay public have the relationship wrong. This error has profoundly affected the understanding of evolutionary biology, the nature of the body's adaptive response, and the testing and assessment of drugs and chemicals, adversely affecting the health of individuals and populations and even national and world economies due to misplaced priorities and extremely wasteful spending. The error originated in the fields of pharmacology and toxicology and, like a highly contagious disease, quickly infiltrated all biological disciplines, as well as government regulatory agencies, including their codified decisions with their non–self-correcting features. This error in judgment on the nature of the dose-response relationship became accepted in the early decades of the 20th century and has been perpetuated to the present time (Calabrese, 2005b, Calabrese, 2005c; Calabrese 2007; Calabrese and Baldwin, 2003a), reinforced by a dominant governmental regulatory and funding culture that strongly influences what scientific ideas will be studied.

This chapter assesses the history of the dose-response relationship and the basis of the error by the scientific community concerning it. The chapter proposes the most basic and appropriate dose-response relationship for the biological sciences along with supportive documentation and a perspective on its broad societal implications.

Historical Antipathies, Rather Than Science, Determined Which Dose–Response Model Would Dominate Biology

The error that determined what has been long considered the fundamental nature of the dose-response relationship was rooted in a scientific version of the Hundred Years' War, that is, the prolonged and bitter conflict between homeopathy and what eventually came to be called "traditional" medicine. To citizens of the late 20th and early 21st centuries, this "medical" conflict would seem to be a minor event, given the overwhelmingly powerful victory of traditional medicine, and therefore not likely to be more than a historical footnote. However, this will be shown not to be the case. As a result of this medical science–based conflict, the basic dose-response relationship—that is, the biphasic dose-response model—got caught in the cross-fire and was victimized because it was a central and highly visible feature of homeopathy.

The linking of the biphasic dose-response relationship to homeopathy was facilitated principally by Hugo Schulz (1853–1932), a professor of pharmacology at the University of Greifswald in northern Germany. Schulz believed that the biphasic dose responses (i.e., low-dose stimulation and high-dose inhibition) he observed in laboratory studies (Schulz, 1888) assessing the effects of chemical disinfectants on yeast metabolism could be broadly generalized and serve as the explanatory principle of homeopathy. Schulz [1923, with English translation by Crump (see Crump, 2003)] emphasized the reproducible nature of his findings in an autobiographic account of the discovery, a perspective that was strongly supported by detailed studies (Branham, 1929) specifically designed to reaffirm and generalize his findings to a wider range of potential antiseptic chemicals. Chester M. Southam and John Erhlich (Southam and Ehrlich, 1943), forestry researchers at the University of Idaho who observed that low doses of extracts from the Red Cedar tree affected the metabolism of multiple fungal strains in a similar biphasic manner, renamed this dose response concept "hormesis" after the Greek word meaning "to excite."

Prior to his intellectually transforming studies with yeasts, Schulz was educated and trained along a traditional biomedical path, with strengths in chemistry and pharmacology. He was also mentored by Eduard Pfluger, one of the founders of modern physiology. However, Schulz was quietly open to homeopathic principles and practices due in large part to an admired and respected family homeopathic physician friend with whom he had a long and intellectually engaged association (Bohme, 1986). At about the time (1882) that Schulz started his career at

Greifswald, research emerged indicating that veratrine, a homeopathic medicine, was a successful treatment for gastroenteritis. Because the causative bacteria had recently been identified and cultured, Schulz (Schulz, 1885) seized the opportunity to assess whether this drug acted via the killing of the bacteria. Extensive tests using a broad range of concentrations revealed that the drug was unable to do so. Although this observation failed to shake Schulz's belief in the efficacy of the drug, it did compel him to conclude that the drug must act via a mechanism other than cell killing.

Several years later when Schulz (Schulz, 1888) observed the biphasic concentration effects of a broad range of chemical disinfectants on yeast metabolism, he came to believe that he had determined how the veratrine might have been effective in the treatment of patients with gastroenteritis. That is, Schulz claimed that at low doses the drug could induce adaptive processes that permitted the person to resist the infection and facilitate recovery. He soon extended this hypothesis to the broader homeopathic field, believing that he had discovered the underlying explanatory principle of homeopathy.

Schulz quickly became a leader within the homeopathic community, devoting the remainder of his professional life to its further study and intellectual expansion. Because Schulz was well known in the pharmacological and medical communities, with numerous publications, as well as active participation on editorial boards of leading professional journals (e.g., *Naunyn-Schmiedeberg's Archives of Pharmacology*) (Starke, 1998), the homeopathic community looked to him to challenge traditional medicine in hopes of legitimizing their medical practices. This also meant that Schulz, his findings, and his interpretations became central in the conflict and the object of considerable criticism by those opposing homeopathic perspectives. The intellectual opposition, that is, traditional medicine, in the form of pharmacology and eventually its scientific offspring toxicology, could not accept Schulz's scientific findings because this would appear as an endorsement of homeopathy.

A careful analysis of Schulz's experimentation (Schulz, 1888) would have revealed that it was not directly relevant to homeopathic medical treatment theory and practice. The vast majority of medical treatments are performed to reduce existing symptoms of illness and prevent their recurrence. This occurs when the individual becomes ill and seeks medical assistance. The homeopathic treatment would normally be expected to be administered after the onset of the illness. In Schulz's work and the overwhelming number of examples of hormesis in the published literature, the investigations did not involve exposures after the onset of disease or chemically induced injury. Even though Schulz believed that his findings were at the core of homeopathic understanding, the scientific community made a critical error in not challenging his interpretation. However, instead it challenged the reliability of Schulz's findings and his dose-response generalization, a decision that would prove to have far-reaching implications for pharmacology and toxicology. Given this strategic, although incorrect decision on how to challenge Schulz, two courses of action emerged: (1) the Schulz biphasic dose-response model (called the

Arndt–Schulz law at the time) had to be marginalized, and (2) a credible alternative had to be formulated, and this becoming the threshold dose-response model, the model on which 20th century clinical pharmacology, toxicology, and risk assessment would be based.

The most notable critic of Schulz was Alfred J. Clark (1885–1941), a highly accomplished pharmacology researcher and scholar, who had considerable influence among academics and government regulators (Verney and Barcroft, 1941; Gaddum, 1962). Nearly 70 years after his death, Clark remains a highly respected figure in pharmacology, with graduate fellowships and a distinguished chair in pharmacology at Edinburgh named in his honor. Clark (Clark, 1933, 1937) was the author of several highly influential, multiedition textbooks that criticized Schulz and his dose-response theories in highly dismissive ways (Calabrese, 2005a) while also linking him with the "extremist" elements within homeopathy (Clark, 1927). In fact, Clark's *Handbook of Pharmacology* was highly regarded, being published as late as 1970, nearly three decades after his death, and influenced several generations of pharmacologists and toxicologists.

Clark's professional successes were due in considerable measure to his careful and objective evaluation of data and his capacity to obtain and integrate massive amounts of complex and technical information in scientifically valid and insightful ways. In the case of his analysis of Schulz, such thoroughness and objectivity were surprisingly below his normally high standards, with a retrospective evaluation (Calabrese, 2005a) revealing that Clark was very selective in his use of the published literature to support his position while failing to report substantial independent findings that supported Schulz's work with yeast and disinfectants (Branham, 1929), as well as his general biphasic dose-response concept (Calabrese and Baldwin, 2000a, Calabrese and Baldwin, 2000b, Calabrese and Baldwin, 2000c, Calabrese and Baldwin, 2000d, Calabrese and Baldwin, 2000e). Of particular note is that Schulz was not in a position to defend himself, given that Clark's criticisms intensified after Schulz entered retirement in the early 1920s, and Schulz died (1932) before the first editions of Clark's two critical books (Clark, 1933, 1937). Furthermore, when the prominent surgical and biomedical researcher August Bier came to his defense, political forces were quickly mobilized to strongly criticize the once-esteemed Bier (Goerig et al., 2000), who had been nominated for the Nobel Prize in Biology and Medicine on multiple occasions, sending a not-so-subtle message to other scientists, even those of considerable achievement and reputation, who might similarly wander from the "party line."

Clark's criticism of homeopathy and Schulz occurred at a time when homeopathic medicine was severely criticized by the so-called Flexner report (Flexner, 1910), which, together with the ongoing efforts by its author, with the backing of the Rockefeller Foundation, over the next two decades effectively led to the closing of the vast majority of homeopathic medical schools in the United States (Berliner, 1985). The final intellectual component of the tipping point regarding the dose-response concept occurred when colleagues of Clark's (Gaddum, 1933; Bliss, 1935) (note that Clark's assistance was acknowledged in the Bliss paper) independently derived the probit dose-response model to account for responses above the so-called toxicology/pharmacology threshold. A critical statistical refinement offered by the esteemed biostatistician R. A. Fischer in an appendix of the Bliss (Bliss, 1935) paper utilized the maximum likelihood estimate to constrain responses to asymptotically approach control-group values in the low-dose zone. In effect, any response below the control group was to be judged as variation, thereby denying the possible biological reality of the J–shaped or inverted–U-shaped dose-response curve. Multiple forces therefore converged and were to control how toxicology and pharmacology considered the dose-response relationship for the next 80 years. Acceptance of the threshold dose-response model would drive the development of these fields, including the selection of animal models, study designs, and risk assessment practices and government regulation.

The exclusion of the hormetic-like dose-response relationship from the mainstream of pharmacology and toxicology during the 20th century was strikingly successful even though there were a substantial number of high-quality research papers supporting the hormetic perspective (Calabrese and Baldwin, 2000a, Calabrese and Baldwin, 2000b). Despite such supportive scientific studies, the hormetic dose-response relationship became marginalized as traditional medicine established its control and directions on the field, transforming the discipline of toxicology in the process. By essentially denying the existence of the biphasic dose-response relationship (Calabrese and Baldwin, 2000c, Calabrese and Baldwin, 2000d, Calabrese and Baldwin, 2000e), 20th century scientific leaders molded toxicology into a high-dose, few-doses discipline. A consideration of the historical development of the reliance of the U.S. National Cancer Institute's (NCI) cancer bioassays on only two doses—the maximum tolerated dose (MTD) and MTD/2—to define the toxicity spectrum illustrates the impact of Clark's dictum on 20th century chronic toxicity and carcinogen evaluations. By truncating the focus of toxicology to above-threshold responses, what did the field miss? Before that question can be answered, it is necessary to establish that biphasic dose responses exist and are reproducible and to discuss their mechanistic foundations and frequency.

The Hormetic Dose-Response Relationship

An important factor in the evaluation of hormetic-like biphasic dose responses is that numerous investigators have reported observations supportive of this dose-response relationship in highly diverse biomedical fields with a notably increased frequency from the mid 1970s and early 1980s to the present. These observations have been associated with various types of technological improvements, including the markedly enhanced capacity to measure lower and lower concentrations of chemicals in various media, thereby permitting toxicological and pharmacological evaluations over a far greater dose range than previously envisioned. It also has been related to major developments in the area of cell culture, including the use of 96-well and higher plates, which permit the assessment of large numbers of chemicals over a broader range of concentrations in a highly cost-effective manner. These advances have been particularly evident in the evaluation of chemically induced immune responses (Calabrese, 2005b), as well as in the assessment of the responses of human tumor cell lines to a wide range of endogenous and exogenous agents (Calabrese, 2005c).

Even though hormetic-like biphasic dose responses have been widely and increasingly reported, it has been common for various biological subdisciplines to use unique descriptors/terms for biphasic dose-response relationships, often specific to each discipline (Table 1). Table 2 provides a historical time line of the citations of some of the main terms used to describe hormetic-like biphasic dose responses based on the Web of Science database. This table documents that the biphasic dose-response concept has shown an increase in citation frequency over the last several decades, that is, long after the dose-response concept had been firmly established and administratively "fixed" within the fields of pharmacology and toxicology and its governmental regulatory agency analogues such as the U.S. Food and Drug Administration, the U.S. Public Health Service, and, since 1970, the U.S. Environmental Protection Agency.

Table 1 Terms Used to Describe Biphasic Dose Responses

U-shaped	Inverted U-shaped
J-shaped	Bidirectional
Biphasic	Hormesis
Dual effects	Arndt–Schulz law
Bimodal	Hueppe's rule
Bitonic	Bell-shaped curve
Pharmacological inversion	Yerkes–Dodson law
Paradoxical effect	Functional antagonism

Table 2 Frequency of Citation by 10-Year Period in Web of Science of Terms That Could Describe Hormesis and Related Terms: 1945–2007

	Frequency of citation						
Term	1945–1954	1955–1964	1965–1974	1975–1984	1985–1994	1995–2004	2005–2007
Bell-shaped curve	0	0	0	1	193	495	128
Bell-shaped dose-response	0	0	0	2	97	252	38
U-shaped dose-response	0	0	0	0	48	195	58
U-shaped curve	0	0	0	1	149	408	145
J-shaped curve	0	0	0	0	21	114	39
Biphasic dose-response	0	0	1	9	182	346	63
Functional antagonism	0	2	0	23	341	1,235	330
Hormesis	1	1	0	10	92	485	247

These terms have been used to describe dose responses with essentially similar quantitative features and general mechanistic strategies. This broad set of descriptor terms for what is believed to the same general dose-response phenomenon has created a significant challenge to the broader biological/biomedical field because this terminological diversity can adversely affect the capacity to discern the general nature of the hormetic-like biphasic dose-response relationship (Calabrese et al., 2007).

The Hormesis Database

To assess hormetic dose responses more systematically and objectively, a database was created in which dose responses had to satisfy rigorous a priori evaluative criteria based on study design, statistical significance, magnitude of the stimulatory response, and reproducibility of findings. Information from this database is now extensive, with more than 8,000 dose responses with evidence of hormesis. Analyses of the database have been used to assess the quantitative features of the hormetic dose-response relationship and their capacity for generalization across biological model, endpoint, and chemical class/physical stressor (Calabrese and Baldwin, 1997; Calabrese and Blain, 2005).

The hormetic dose-response relationship is similar to the threshold dose-response relationship for responses that exceed the toxicological/pharmacological threshold. However, it is below this threshold where the two dose-response models differ. In the case of the threshold dose-response model, there is the assumption that there is no significant treatment-related effect below the threshold, with responses predicted to randomly bounce above and below the control-group value. With respect

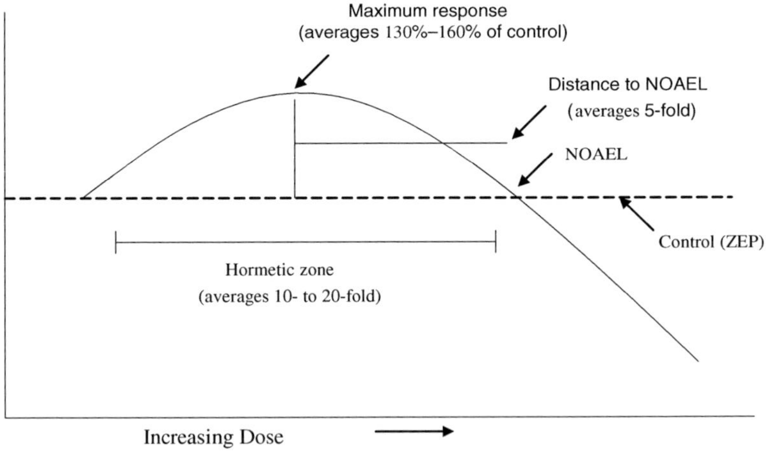

Fig. 1 Dose-response curve showing the quantitative features of hormesis. NOAEL, no observed adverse effect level; ZEP, zero equivalent point (ZEP) (i.e., value equal to the control value)

to the hormetic dose-response model, there is the expectation that responses will nonrandomly increase above that of the control group starting immediately below the threshold response. Analysis of data from the hormesis database (Calabrese and Blain, 2005) indicates that the magnitude of the stimulatory response is characteristically modest, almost always less than twice that of the control group, with the strong majority of maximum stimulatory responses being within the range of 30% to 60% greater than control values (Fig. 1). The width of the stimulatory response is generally within a range extending from immediately below the threshold to about 1/20 of the threshold value. However, the stimulatory range may vary appreciably, with about 2% of dose responses in the hormetic database having a stimulatory width of greater than 1,000-fold. Figure 2 displays a broad range of examples of hormetic dose-response relationships. These examples illustrate that hormetic

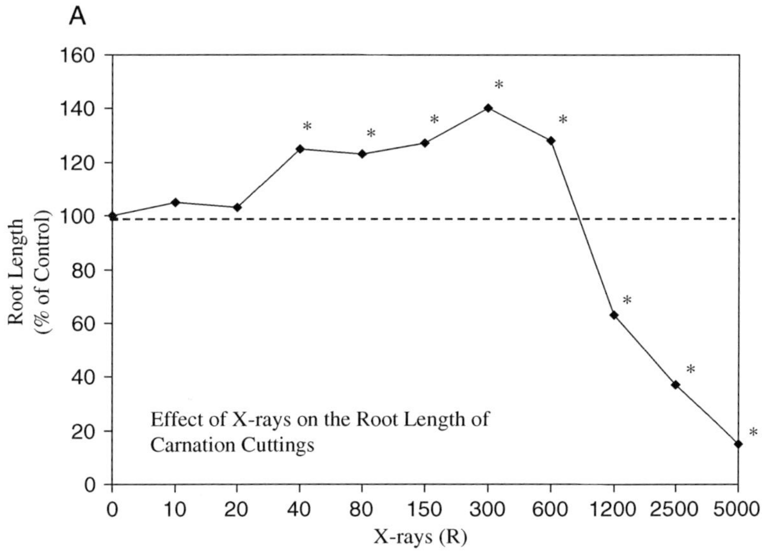

Fig. 2 Selected examples of hormesis, reflecting a broad range of biological models, endpoints, and chemicals/physical stressor agents. **A.** Effect of X-rays on the root length of carnation cuttings (Bors and Zimmer, 1970). **B.** Effect on primary astrocyte cultures with the MTT [3-(4,5-dimethylthiazol-2-yl)-2,5-diphenyl tetrazolium bromide assay] (Cookson et al., 1995). **C.** Effect of gamma rays on the lifespan of female house crickets. (Hunter and Krithayakiern, 1971). **D.** Effect of acridine on reproductive performance in *Daphnia* (Parkhurst et al., 1981). **E.** Effect of lead and copper on the survival of springtails (Sandifer and Hopkin, 1997). **F.** Effect of dexamethasone on cell growth and viability of cultured human retinal pigment epithelial (RPE) cells (Wu et al., 2002). **G.** 2-Methyl-4-chlorophenoxyacetic acid (MCPA) and oat shoot growth (Wiedman and Appleby, 1972). **H.** Cadmium and aquatic plant nitrate reductase activity (Rai et al., 1998). **I.** Aluminum and mouse blood gamma-aminolevulinic acid activity (Vieira et al., 2000). **J.** Alcohol and rat serum levels (Cicero and Badger, 1977). **K.** Cyclopentyladenosine (CPA) and porcine coronary artery (Merkel et al., 1992). **L.** Arsenic and human lymphocyte DNA synthesis (Meng 1993). **M.** Effect of streptomycin on mortality in mice (Welch et al., 1946). **N.** Modulation of fibroblast proliferation of oxygen free radicals (Murrell et al., 1990)

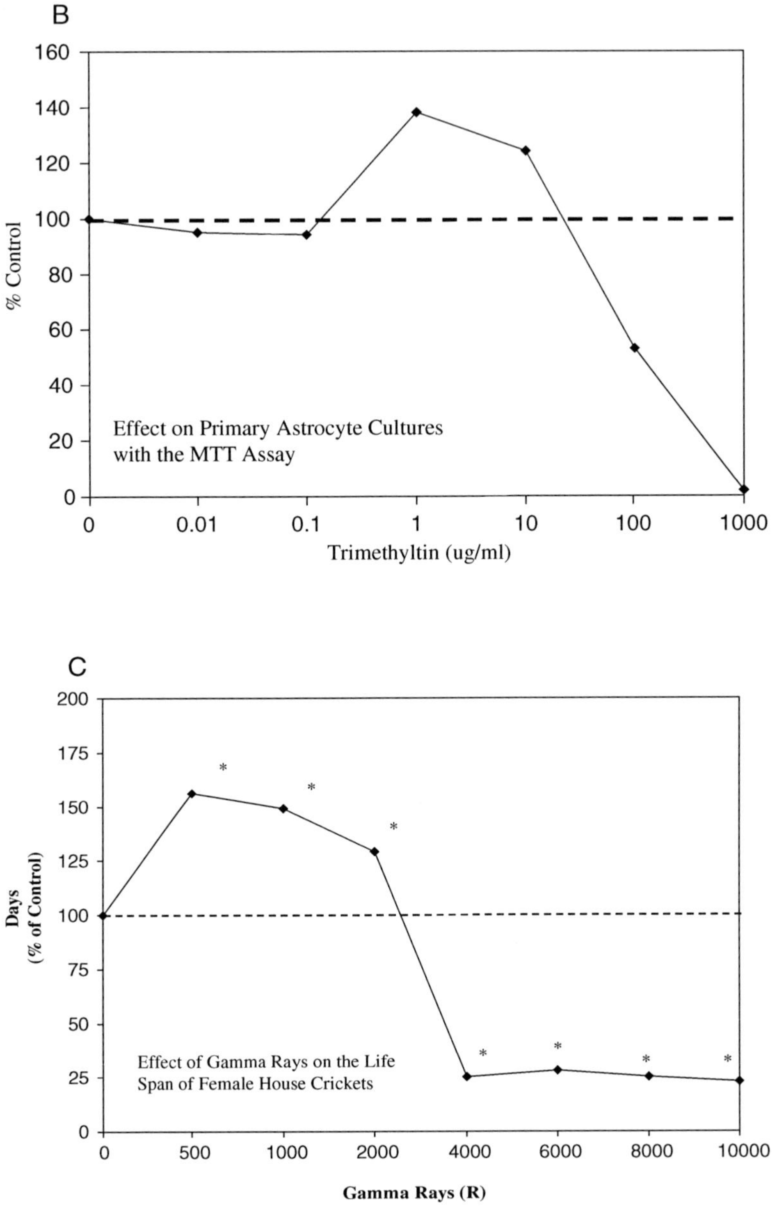

Fig. 2 (continued)

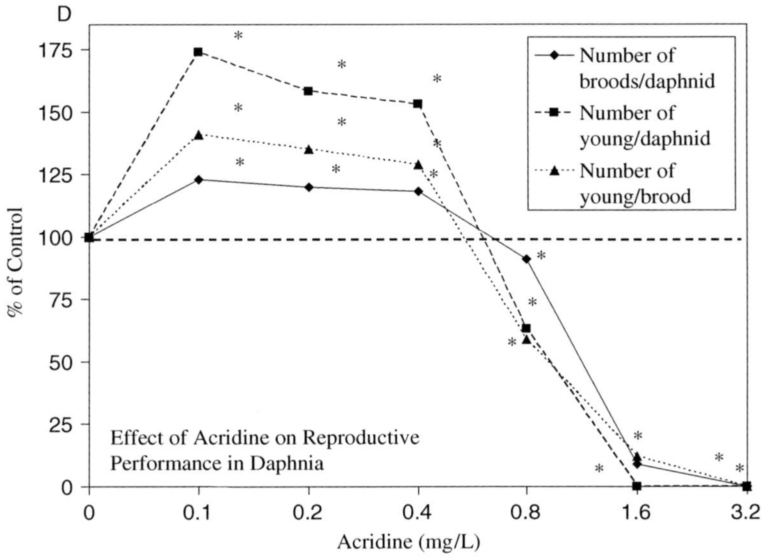

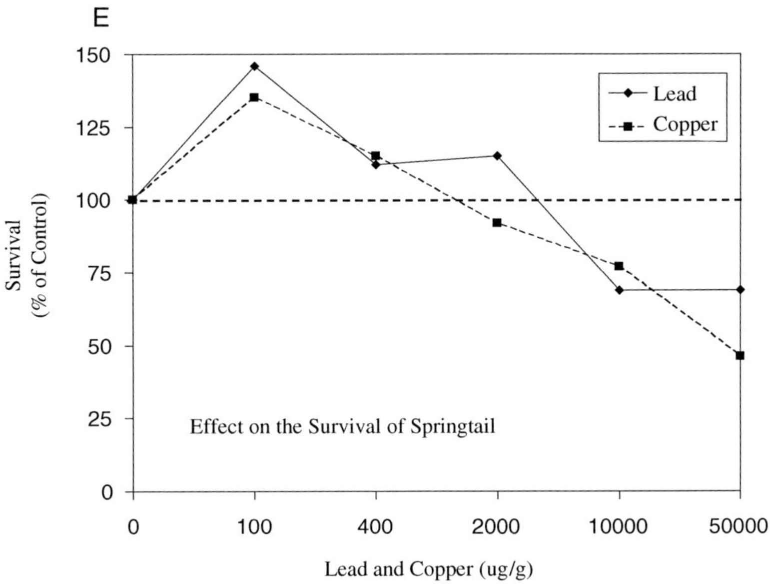

Fig. 2 (continued)

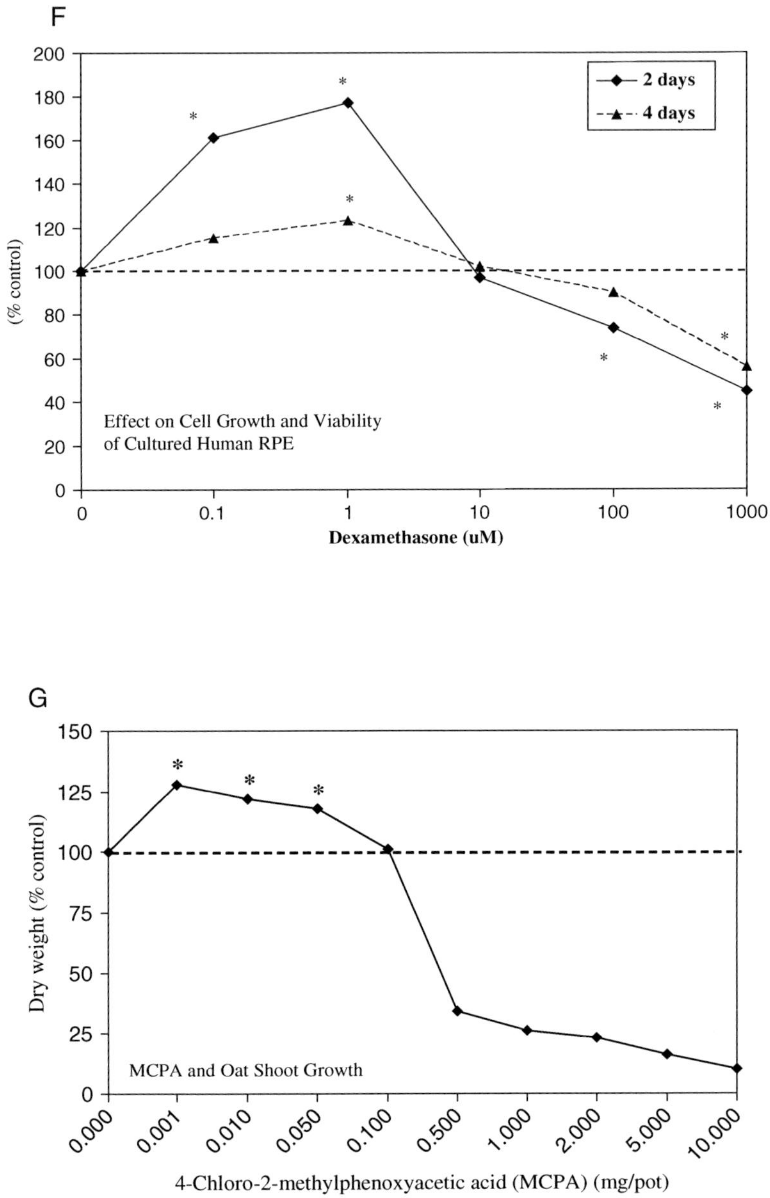

Fig. 2 (continued)

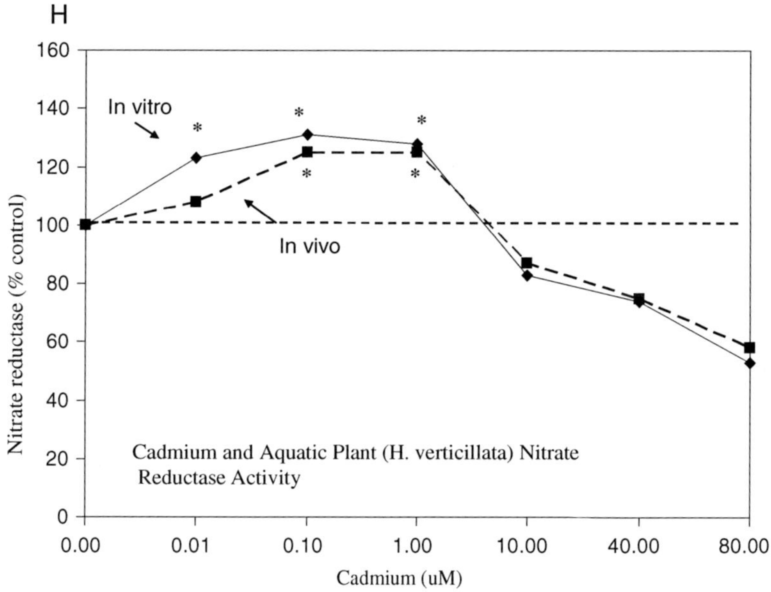

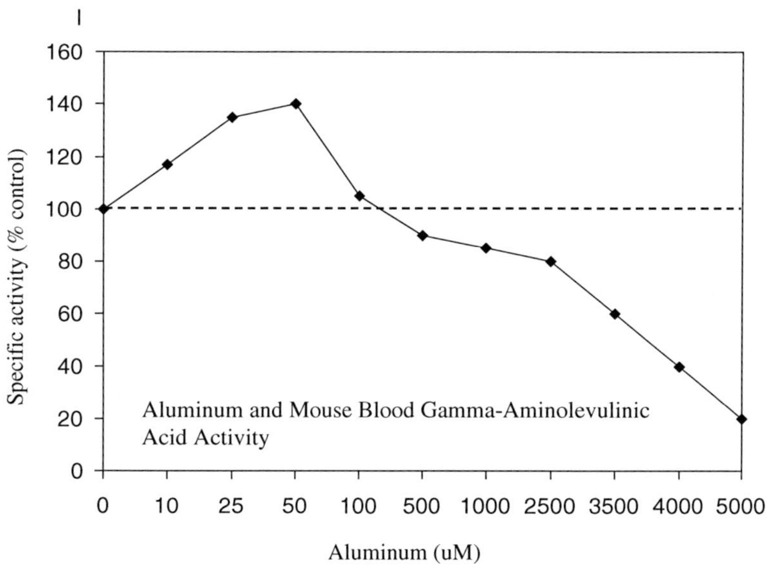

Fig. 2 (continued)

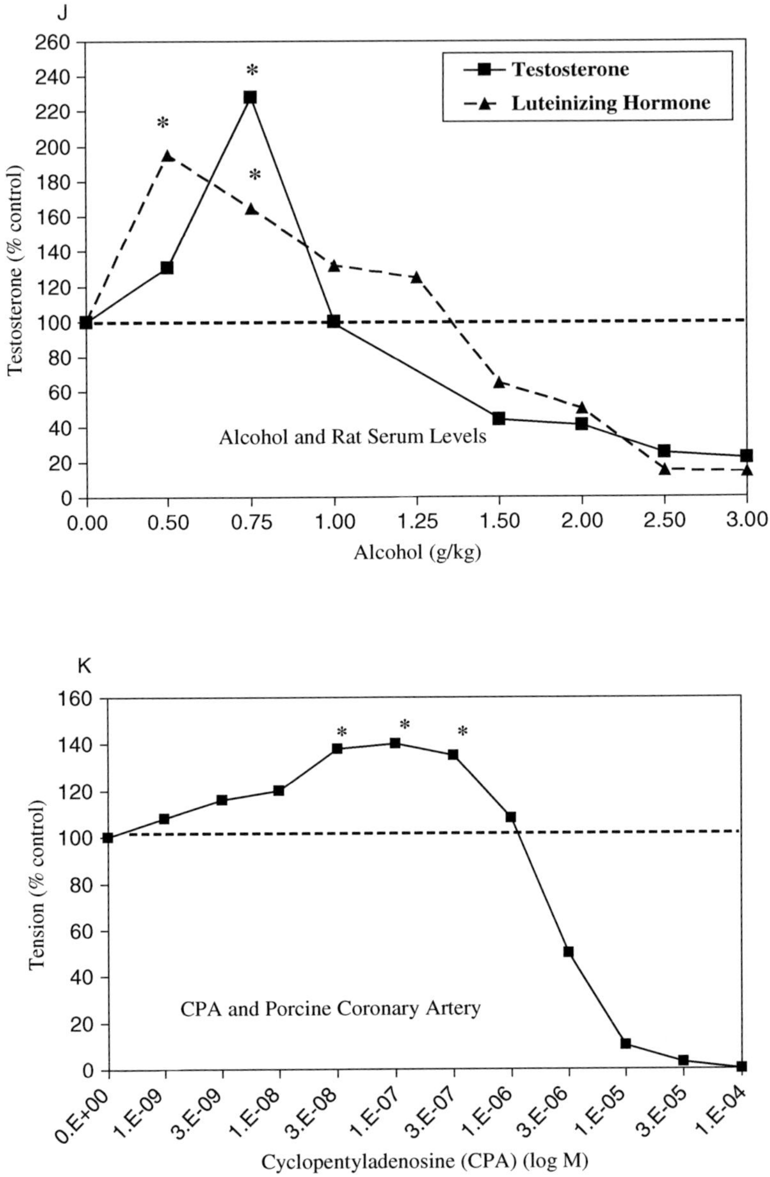

Fig. 2 (continued)

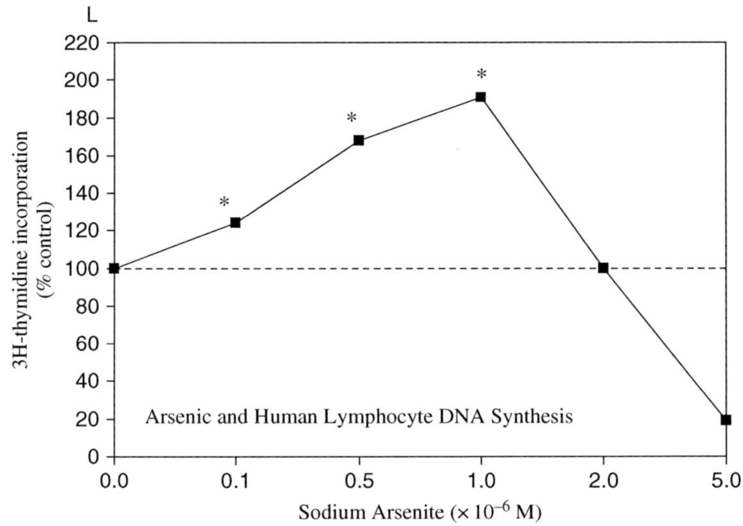

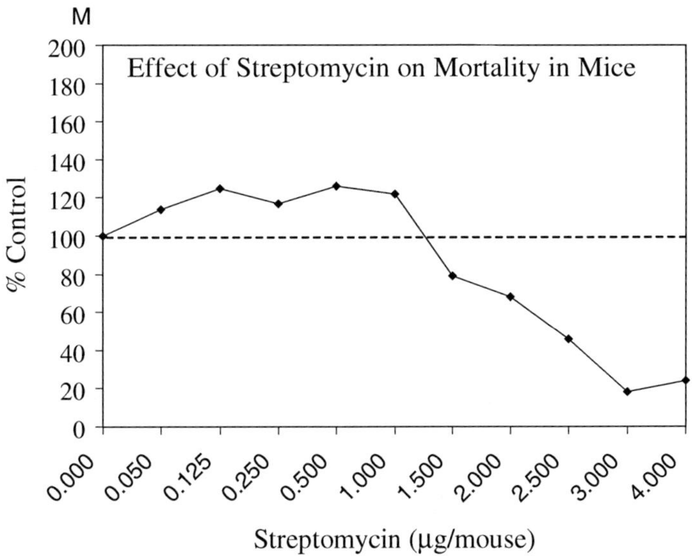

Fig. 2 (continued)

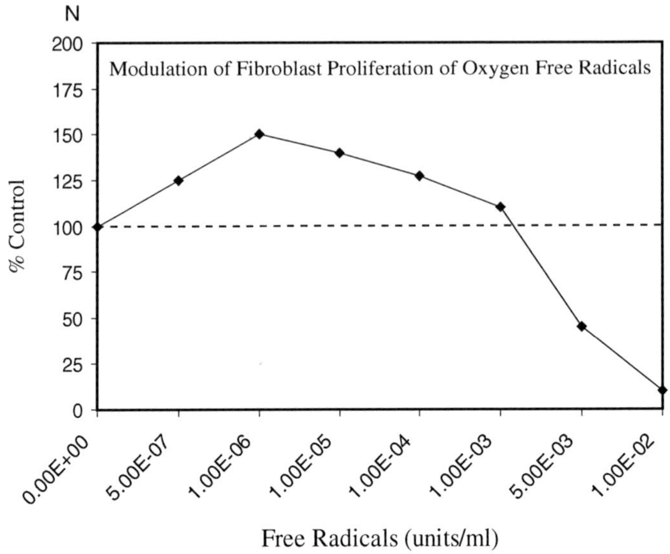

Fig. 2 (continued)

effects occur across a wide range of biological models, endpoints, and chemical classes. Detailed assessment of the hormetic database indicates that the hormetic dose-response relationship is both highly generalizable and conserved within an evolutionary framework.

The Frequency of Hormesis in Toxicology and Pharmacology

A second hormetic database was created to assess a limitation in the aforementioned database. That is, its lack of a priori entry criteria prevented the capacity to estimate the frequency of hormesis in the toxicological and pharmacological literature. Even though there was considerable evidence demonstrating the occurrence of hormetic dose-response relationships and their biological and statistical features, the database could not address the question of whether hormetic responses would be expected to occur in 1% or 50% of the cases of properly designed experiments. This is important, especially for regulatory agencies that may have different priorities for responses with a low frequency than for those that are common and highly generalizable. Such determinations may affect strategies for how regulatory agencies manage such chemicals within a standard setting framework. If hormetic effects are of low frequency (e.g., occur in less than 1% of studies), they may be treated on a case-by-case basis. If they occur with fairly high frequency, then general procedures would likely be developed for their systemic evaluation. In the case of the second hormetic database, approximately

21,000 articles were assessed for their capacity to satisfy initial a priori entry criteria; those satisfying such criteria were then subjected to a priori evaluative criteria. This assessment revealed that hormetic effects occurred in nearly 40% of the cases that satisfied both the entry and evaluative criteria (Calabrese and Baldwin, 2001). These findings provided the first general frequency estimate of hormetic dose responses in the toxicological and pharmacological literature. This database was used subsequently to test the capacity of the hormetic and threshold dose-response models to predict responses below toxicological/pharmacological thresholds. In this specific assessment, hormetic dose responses occurred far more commonly (i.e., 2.5:1) than the threshold dose-response model predicted (Calabrese and Baldwin, 2003a). These hormetic findings were extended via the use of a U.S. NCI database on the effects of nearly 2,200 potential antitumor drugs on 13 strains of yeast (i.e., 57,000 concentration responses) with well-characterized genetic alterations relating to DNA repair and cell cycle control (Calabrese et al., 2006). These findings add strong support to the hypothesis that reproducible stimulatory responses occur below traditional toxicological/pharmacological thresholds and that these responses, although quite common, are poorly predicted by the threshold model, whereas the hormetic model predicts such responses with considerable accuracy.

Comprehensive assessments of hormetic responses have been published in the areas of immunology (Calabrese, 2005b), human tumor cell lines (Calabrese, 2005c), and neuroscience, including neuroprotection (Calabrese, 2008a), neurite outgrowths (Calabrese, 2008b), memory (Calabrese, 2008c), pain (Calabrese, 2008d), stress responses (Calabrese, 2008e), anxiolytic drugs (Calabrese, 2008f), antiseizure medications (Calabrese, 2008g), and stroke (Calabrese, 2008 h), and detailed critical assessments are available in other areas as well, including chemotherapeutics (Calabrese, 2003a), apoptosis (Calabrese, 2001a), cellular migration behavior (Calabrese, 2001b), environmental contaminants (i.e., inorganics) (Calabrese, 2003b), and numerous endogenous agonists, including dopamine (Calabrese, 2001c), nitric oxide (Calabrese, 2001d), estrogen (Calabrese, 2001e), testosterone (Calabrese, 2001f), serotonin (Calabrese, 2001g), opioids (Calabrese, 2001h), and adrenergic agents (Calabrese, 2001i). This substantial body of literature indicates that hormesis is highly generalizable, being independent of biological model, endpoint, and stressor. These findings indicate that the threshold models and models that are linear at low doses often fail to accurately predict low-dose responses.

Implications of Hormesis

Detailed evaluations of the hormetic database indicate that the hormesis concept may have significant impacts on multiple areas, including biological concepts, toxicological/pharmacological principles, environmental risk assessment theory and practices, clinical medicine, and agricultural/industrial applications. These are now discussed.

Impact on Biological Concepts

Hormesis Measures Performance

The low-dose stimulation as observed in toxicologically based hormetic dose-response relationships reflects *biological performance*, whereas inhibitory responses occurring at higher doses typically describe toxicity. This general scheme is also the case within a pharmacological framework in which a similar stimulatory response occurs typically at doses below the threshold, whereas either inhibition or a return toward control values occurs as the dose increases due to toxicity, receptor desensitization, or other factors.

Toxicology has long been dominated by an emphasis on very high doses and the assessment of toxic responses. In the case of toxicity, there is an expansive potential for an increase in the expression of injury as the dose increases beyond the threshold. This may be seen in the release of tissue enzymes into the serum as in the case of biomarkers of liver damage or the number of tumors per animal in cancer bioassays. The potential for the system to display biological performance as seen with the hormetic dose response has been routinely missed because below-threshold responses have not been systematically studied. Recognizing that the hormetic stimulatory response is a manifestation of biological performance is a novel conceptual interpretation. Biological performance responses occur at all levels of biological organization, conform to the constraints of biological plasticity, and optimize system function. There are numerous endpoints that express biological performance, including cell proliferation, growth, longevity, strength, disease resistance, increases in cognition, and others. Thus, the hormetic dose response expands the dose-response concept, with the expectation that there is both an above-threshold toxicity feature and a below-toxicity, stimulatory component that describes biological performance.

There are conditions when the low-dose performance stimulation may not be beneficial to the individual. This may occur in the case of enhanced cellular proliferation causing organ enlargement such as the prostate (Chueh et al., 2001), enhanced risk of a detached retina via retinal epithelial cell proliferation (Wu et al., 2002), or the proliferation of harmful microorganisms (Randall et al., 1947; Garrod, 1951). Nonetheless, the concept of hormesis as an expression of biological performance in these cases remains valid, even if it occurs in an infectivity circumstance or when a tissue can be biologically "tricked" into a response that may be harmful to the organism. The performance response remains one that is of limited magnitude, is constrained by plasticity, and optimizes a response function.

Hormesis Provides Quantitative Estimates of Biological Plasticity

Because the performance function is highly generalizable, with consistent quantitative features, the hormetic dose response may represent a general estimate of the magnitude of biological plasticity across biological systems and possibly account for the marked constraints in the magnitude of hormetic stimulatory effects. If this were the case, it would represent a significant unifying biological concept. It suggests that the low-dose performance stimulation, that is, the hormetic stimulatory response, may provide the quantification of one major type of biological plasticity.

Adaptive Response/Preconditioning: Manifestations of Hormesis

Over the last several decades there has been considerable research on the adaptive response in radiation and chemical toxicology. The adaptive response occurs when a prior low dose of a toxic agent or stress condition enhances the capacity of the affected cell, organ, or organism to resist the toxicity of a subsequent and more massive exposure to the same or similar agent or stress condition (Calabrese, 2008i). A similar type of adaptive response was named preconditioning after investigators observed that a hypoxic stress administered 24 hours prior to a massive myocardial infarction in dogs reduced cardiac damage by nearly 80% (Murry et al., 1986). Both the adaptive response and preconditioning concepts describe a similar temporal process in which a low prior dose of a stressor agent upregulates a cascade of molecular events that results in the temporary protection against a subsequent substantial threat. The key connection of the adaptive/preconditioning response to the hormesis concept is found in a detailed evaluation of the "adapting" or "preconditioning" doses. That is, prior dosing displays a dose-response optimum that maximizes the subsequent protective effect. Lower and higher doses display a drop-off of the protection response. Higher adapting or preconditioning doses may act to further exaggerate the toxicity of the more massive subsequent exposure. If this relationship is plotted, it represents a biphasic dose response with quantitative features that are consistent with the hormetic dose response.

The relationship of hormesis to the adaptive response was explored by Davies et al. (Davies et al., 1995), who defined the optimal condition for a transient hydrogen peroxide adaptation as measured by cell viability in the yeast model *Saccharomyces cerevisiae*. In a critical first step, the authors determined the effects of hydrogen peroxide on cell proliferation employing up to nine concentrations. A hormetic-like biphasic dose-response relationship was reported in which low hydrogen peroxide concentrations (0.4 mM or less) enhanced cell colony growth by approximately 30%. The hydrogen peroxide–induced toxicity started to occur between 0.5 and 0.8 mM. Based on these findings, an adapting or preconditioning dose was selected to be one in the low-dose stimulatory/hormetic zone. The yeast cell treatments that received the adapting dose in the hormetic zone followed by the challenging (i.e., cell killing) dose not only showed the adaptive response, but also displayed a percentage viability that exceeded the original control value by approximately 20% to 50%.

Hormesis as an Expression of Allometry

A significant concept in the biological sciences is that of allometry, which provides a quantitative integration of numerous biological parameters as a function of body weight and/or surface area (Calder, 1996). Allometric relationships are particularly important for the toxicological and pharmacological sciences because they provide a biologically based biomathematical framework for estimating the responses of drug treatments both on an interindividual basis and extrapolated across species. An important observation is that hormetic dose-response relationships can be modeled allometrically, and these relationships are consistent within and between species. The hormetic response represents a similar proportional increase to a normalized control group independent of species, thereby providing the basis for the allometric relationship. This observation strongly supports the generalizability of the hormetic dose response across species for human endpoints and integrates this concept within an evolutionary framework.

Toxicological/Pharmacological Implications

Factors Affecting the Recognition of Hormetic Dose-Response Relationships

Use of Multiple Terms

Many terms have been used to describe the hormetic dose-response relationship (Table 1). The use of such a wide range of terms, many specific to biological subdisciplines, for the same quantitative features of the dose-response relationship has created conceptual confusion on the nature of the relationship in the low-dose zone. One significant contributory factor to the use of such a wide range of terms for the same apparent dose-response concept is the progressive specialization within the sciences, which reduces communication between specialties.

Modest Stimulation and Historically Weak Study Designs

Further contributing to the difficulty in recognizing the occurrence, generalizability, and reproducibility of the hormetic-like biphasic dose-response relationship is that most hormetic dose-response relationships are characterized by a modest stimulation (30% to 60%) in the below-threshold zone, a feature that is its most distinguishing characteristic (Calabrese and Blain, 2005). Given the modest magnitude of the hormetic stimulatory response, it can be difficult to verify when studies have only few doses that are intended to document toxicity and estimate the toxic threshold.

Control Group: High Variation

The use of biological models with high background variability is problematic in the evaluation of hormetic hypotheses. The presence of such variability places heightened demands on sample size to increase statistical power to evaluate possible treatment effects.

Low Background Disease Incidence

The field of toxicology adopted the use of biological models with very low disease incidence to maximize statistical power while using as few subjects as possible to reduce financial expenses associated with the conduct of experiments. This set of study design challenges still exists, as evidenced in the standard testing protocols required by most governmental regulatory agencies. When a control group displays a very low background disease incidence, it is essentially impossible, at least in a practical sense, to assess hormetic dose-response hypotheses. This testing strategy, which has historically governed hazard assessment, indicates that hazard assessment goals and practices lack the capacity to assess the presence or absence of hormetic-like biphasic dose-response relationships. The use of control groups with such negligible disease incidence reinforces the belief that the threshold dose-response model is valid and appropriate for extrapolation in the low-dose zone, an assumption that has been discredited (Calabrese and Baldwin, 2003a; Calabrese and Baldwin, 2001; Calabrese et al., 2006).

Lack of Temporal Component

Hormetic dose responses may be difficult to discern because they often require a time component that is typically lacking in standard hazard assessments. Because hormetic dose responses represent a modest overcompensation stimulation following a disruption in homeostasis (i.e., toxicity), it is necessary to document the biological responsiveness over time to assess the effects of the chemical or physical stressor. The hormetic dose response can therefore be easily missed if the biological model is not tested over the proper range of doses and within the appropriate temporal framework.

Summary

In the chronic bioassays required by regulatory agencies, there is little to no opportunity to assess hormetic effects because they use too few doses, very high doses, and animal models with low background disease incidence. Failure to consider the possibility of a hormetic dose-response relationship has significant implications, preventing the identification of possible harmful or beneficial effects in the below-threshold zone.

Chemical Potency and Hormesis

Pharmaceutical agents that affect the same endpoint may often display profoundly different potencies. However, despite the fact that an agent may be far more potent than another agent for the same endpoint, the quantitative features of the hormetic response are remarkably similar (Calabrese, 2008d) (Fig. 3). As suggested earlier,

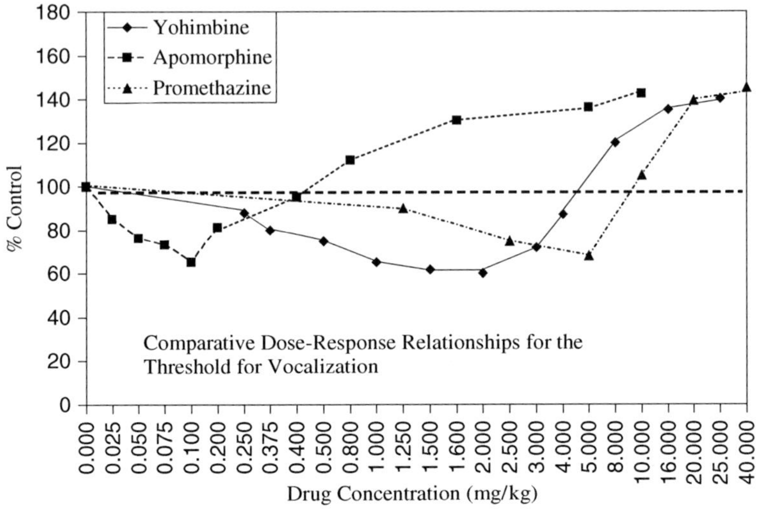

Fig. 3 The J-shaped dose-response relationships for yohimbine, apomorphine, and promethazine for pain. Comparison of Figs. 1, 2, and 4 of Paalzow and Paalzow (1983a), Paalzow and Paalzow (1983b), and Paalzow and Paalzow (1985), respectively

stimulatory responses are similarly limited by the constraints imposed on biological systems by their inherent plasticity. Potency therefore does not affect the quantitative features of the hormetic dose response.

Hormesis: A Novel Concept of Synergy/Potentiation

There is considerable research on chemical interactions concerning toxicity, that is, the above-threshold side of the dose-response relationship (Calabrese, 1991). Regulatory agencies have written policy statements for how to assess interactions that are additive or synergistic (U.S. EPA, 1986). However, they have not addressed the issue of chemical interactions on the performance side of the dose-response relationship (i.e., below the threshold) (Calabrese, 2008j). The distinction between the toxicity and the performance parts of the dose-response relationship with respect to chemical interactions also has not been considered in the assessment of the effects of pharmaceuticals on human health. In the case of chemical interactions and hormesis/performance, the dose response in the low-dose stimulatory zone is constrained by the limits imposed by biological plasticity. Increases in response would not be expected to exceed the modest stimulation of only 30% to 60% rather than the multifold possible increases seen in the toxicity side of the dose response relationship. In a practical sense, hormesis-related synergy would be observed as the reduction in dose needed to achieve a near-maximum response that would be constrained by plasticity. This was reported by Flood et al. (Flood et al., 1983, 1985), who assessed the effects of drugs on cognition in rodent models. They combined exposures of agents that could individually maximally enhance cognition by about 50%. When these drugs were combined in various ways, the maximum response was not increased, but the amount of agent needed to achieve the maximum, although modest, response was profoundly reduced for each of the agents used in the drug combination experiments. Thus, synergy in these performance-oriented experiments was most clearly observed at the level of dose rather than at the level of response. A practical implication of these findings is that even though the absolute degree of cognition was not enhanced in drug combinational studies beyond what was achieved with a singly acting agent, the modest maximum response could be achieved by using far lower quantities of the several combined drugs, thereby markedly reducing the likelihood of undesirable drug side effects.

Even though the chemical interaction concept was observed best on the dose side in the assessment of hormesis/performance, it could also be seen on the response magnitude side. However, any response synergy would be difficult to observe because it would have to occur within the 30% to 60% maximum response range. This suggests that response-side examples of synergy within a hormetic setting can be addressed only within experimental settings in which the control group has very low background variability.

The implications of the constraints of biological plasticity on chemical interactions are important for the pharmaceutical industry. Drug treatments, whether administered singly or in combination, are not likely to achieve a response greater than the hormetic maximum. Thus, if a drug increases memory by 30% to 60% in an Alzheimer's disease patient, it should not be expected that a drug combination would exceed this value.

Interindividual Variation and Hormesis

The hormetic/performance stimulatory response begins immediately below the toxic threshold. There is often less than a factor of 5 to 10 between the optimal dose for performance and the onset and occurrence of toxicity. Because humans often display a range of interindividual variation between 10- and 20-fold and sometimes even greater (Calabrese, 1985), estimated "optimal" doses of drugs for the so-called average person are likely to display a range of possible responses in a heterogeneous population, including being highly effective, that is, achieving the optimal zone, but also exposures at which the dose misses the optimal zone on either the low or high side, resulting in little or no treatment effect or possible toxicity, respectively. The overlapping of optimal performance and toxicity zones due to patterns of interindividual variation in response is common in clinical practice. This requires continual fine-tuning to optimize the patient treatment dose. It would be far less challenging for the clinician if the goal were to kill harmful microbes or tumor cells. In this case, the physician would be directing attention to the toxicity side of the threshold response. This phenomenon also has important implications for the interpretation of epidemiological studies, as discussed next.

Epidemiology and Hormesis

Dose-response relationships may be markedly affected by the degree of heterogeneity within the population. This would be a particular consideration for epidemiological investigations in which there is considerable ethnic, age, and health variability. To assess the influence of heterogeneity on the overall integrated population-based dose-response relationship, we conducted a series of simulations in which a number of subgroups were identified and given a specific dose-response relationship characteristic that could be hormetic, threshold, or linearity, and we took differential proportions of the total population. This exercise demonstrated that by altering any of the foregoing parameters, one could significantly change the overall shape of the population-based dose-response relationship. The final integrated population-based shape of the dose-response relationship could be readily made to become linear, threshold, biphasic, or multiphasic.

This exercise indicated that epidemiological evaluations could be problematic in the assessment of a hormetic or any other type of dose-response relationship. The occurrence of the intergroup dose-response variability and its differential proportion within the population may create a blended dose-response relationship that could mask the dose-response dynamic that occurs at each subgroup level. Recognition of this possible complexity in assessing the nature of the dose response, especially in the low-dose zone, is an important consideration affecting data interpretation.

Hormesis and Medicine

Hormesis has profound implications for the field of medicine because it defines the qualitative and quantitative features of the dose response. We now briefly describe a broad spectrum of medical applications.

Low-Dose Stimulation of Tumor Cells

The hormetic dose-response relationship predicts that antitumor agents may enhance the proliferation of tumor cells in the low-dose zone. This prediction was confirmed in an extensive review of the effects of antitumor agents on the proliferation of human tumor cell lines (Calabrese, 2005c). Hormetic-like biphasic dose responses were reported in 136 human tumor cell lines from more than 30 tissue types for more than120 agents. Although the mechanisms were often different, being specific for each tumor type, the shape of the dose-response relationships is consistent with the hormetic biphasic model. Even though the endpoint measurement for response varied from 1 hour to 21 days, depending on the agent and the tumor cell line, a hormetic biphasic dose-response relationship was consistently reported.

These findings suggest that many antitumor agents have the potential to enhance tumor cell proliferation in patients. This situation would be of particular concern for agents with long biological half-lives. For example, the chemotherapeutic antitumor agent suramin has a human half-life between 30 and 40 days and has the capacity to induce tumor cell proliferation in a biphasic dose-response fashion (Foekens et al., 1992).

Low-Dose Stimulation of Microbes by Antibiotics

Similar hormetic-like biphasic dose responses have been reported for a wide range of chemicals on colony growths of bacteria, fungi, yeasts, and algae (Calabrese and Baldwin, 2000a). Similar findings were reported with synthetic antibiotics soon after their discovery in the early to mid 1940s using in vitro (Garrod 1951; Miller et al., 1945; Ungar and Muggleton, 1946) and in vivo studies (Randall et al., 1947; Welch et al., 1946) (Fig. 2m). In addition, Foley and Winter (Foley and Winter, 1949) reported that penicillin increased the mortality of chick embryos inoculated with *Candida albicans*. The possibility that low concentrations of antibiotics may have contributed to reports of enhanced patient morbidity and mortality was raised in various reports (Garrod, 1951). It is well known that various antiviral drugs can facilitate the proliferation of a broad range of viruses (Lee et al., 1999; Nyberg et al., 2004). However, the clinical implications of these findings remain to be explored.

Anxiolytic Drugs

The hormetic dose response describes the dose-response features of anxiolytic agents regardless of which receptor pathway mediates the response. In the strong majority of cases, low doses reduce anxiety in animal models, whereas higher doses increase anxiety (Calabrese, 2008f; Melchior and Ritzmann, 1994) (Fig. 4). Such

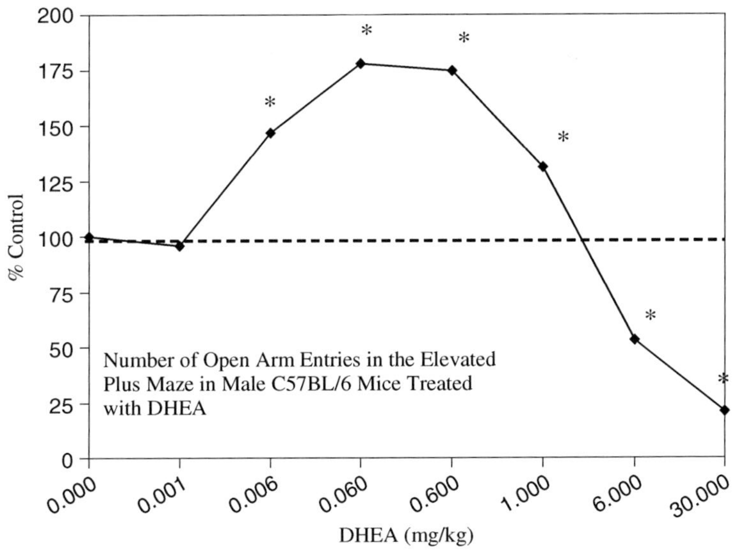

Fig. 4 Number of open arm entries in the elevated plus maze in male C57BL/6 mice treated with dehydroepiandrosterone (DHEA). *Significantly different from controls at $P < 0.05$ (Melchior and Ritzmann, 1994)

findings are also consistent across the wide range of experimental behavioral protocols (e.g., elevated plus maze, social interaction test, open-field test, light-dark test, hole board test) that are routinely used to assess anxiolytic drugs.

Antiseizure Drugs

Low doses of antiseizure drugs typically increase the threshold for the initiation of seizure responses, whereas at higher doses the onset of seizures is facilitated as the threshold for response is decreased. In screening studies for possible antiseizure drugs, animal models are routinely administered standard seizure-inducing agents, such as pentylenetetrazol (PTZ), that cause a reliable seizure response at specific doses, depending on the animal model. Possible antiseizure drugs are those that demonstrate the capacity to increase the threshold for the induction of seizures by agents such as PTZ, that is, make it more difficult for a seizure to occur. An example of this hormetic phenomenon was reported (Honar et al., 2004) with respect to morphine (Fig. 5).

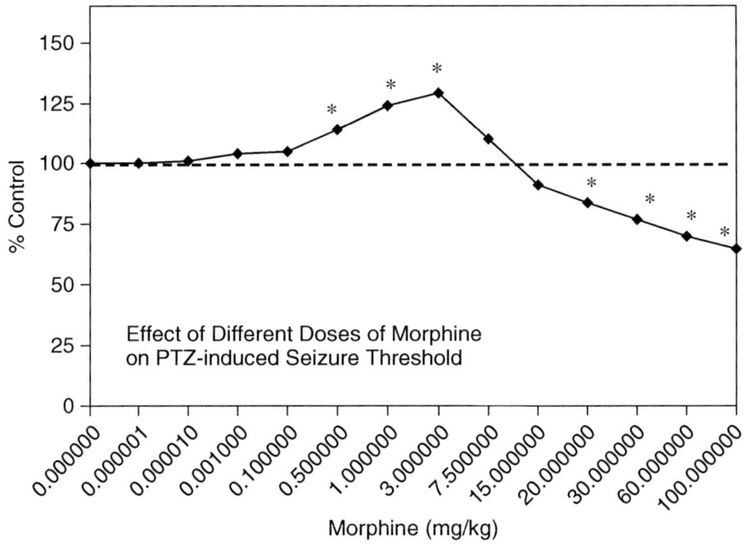

Fig. 5 Effect of different doses of morphine on pentylenetetrazol (PTZ)-induced seizure threshold. *Significantly different from controls at $P < 0.05$ (Honar et al., 2004)

Memory-Enhancing Drugs

Numerous drugs have shown a capacity to enhance learning and memory in animal models, starting with the seminal work of McGaugh and Petrinovich (McGaugh and Petrinovich, 1965) with the anticholinesterase agent physostigmine. In general, such memory-enhancing drugs exhibit a U-shaped dose-response relationship regardless of the model and the specific learning or memory endpoint considered or whether such drugs are given in combination (Flood et al., 1983, 1985). This generalized pattern of hormetic-like biphasic responses is seen with all Alzheimer's disease drugs approved by the U.S. Food and Drug Administration (FDA) (Calabrese, 2008c; Wise and Lichtman, 2007).

Stroke Medications

Nearly three dozen possible stroke and acute brain injury medications reduce damage via hormetic mechanisms that show the standard biphasic dose-response relationship (Calabrese, 2008h). Such hormetic-like biphasic dose-response relationships have employed a diverse set of stroke and injury models, as well as agents that act via specific mechanisms and stages of the injury prevention or tissue recovery process.

Osteoporosis

Bisphosphonates, which are widely used in the treatment of osteoporosis, follow the hormetic dose-response relationship. Detailed in vitro animal model and human investigations have demonstrated such dose responses for osteoblast formation (Giuliani et al., 1998) (Fig. 6) at drug doses that are generally equivalent to those used in the management of humans with osteoporosis (Liberman et al., 1995; Rossini et al., 1994).

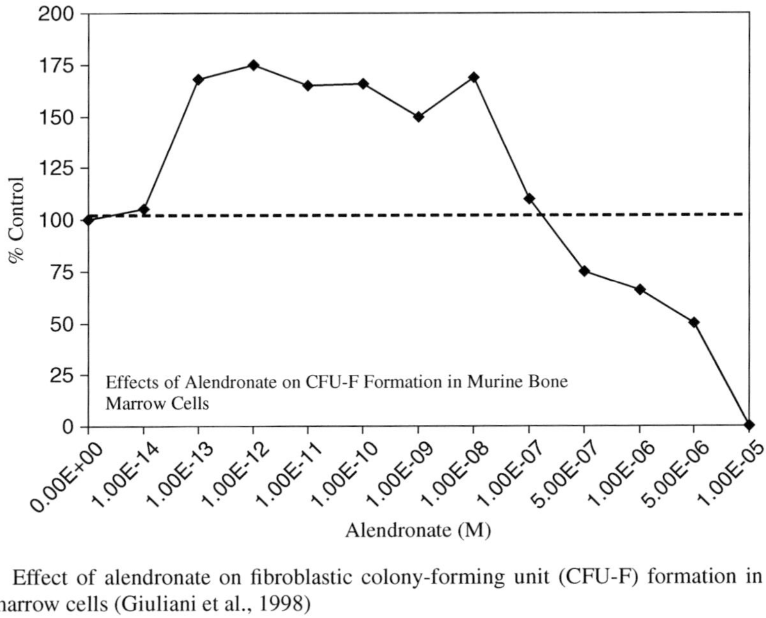

Fig. 6 Effect of alendronate on fibroblastic colony-forming unit (CFU-F) formation in murine bone marrow cells (Giuliani et al., 1998)

Hair Growth

Drugs such as minoxidil that enhances hair growth also do so in a manner that is consistent with the hormetic dose-response relationship (Boyera et al., 1997).

Pulmonary Hypertension

Epidemiological investigations have associated the long-term administration of Prozac with a reduced risk of developing pulmonary hypertension in adults while increasing the risk of this disease in the fetus. Follow-up research with Sprague-Dawley rats revealed that this drug reduces the occurrence of pulmonary arterial smooth muscle cell proliferation in adult female rats in a dose-dependent manner (Fornaro et al., 2007). However, Prozac induced a hormetic-like low-dose stimulation and a high-dose inhibition for the same endpoint in the fetal rat. These findings were consistent with the protective effect seen in adult humans and the increased risk of pulmonary hypertension in the fetus. The findings were especially interesting in that a low dose induced a harmful effect, whereas the higher exposure diminished the risk of pulmonary hypertension.

Fibrotic Diseases (e.g., Dupuytren's Contracture)

Pathological fibrotic conditions are associated with the presence of fibroblasts at high cell density. Many of the biochemical and ultrastructural features of fibrosis are thought to be secondary to the increase in fibroblast density. According to Murrell et al. (Murrell et al., 1990), the progression of Dupuytren's contracture, a fibrotic condition of the hand associated with microvascular ischemia, occurs by exposure to oxygen free radicals that can stimulate and inhibit the proliferation of cultured human fibroblasts in a hormetic-like biphasic manner (Fig. 2n). Prolonged stimulation in the low-dose zone may, therefore, promote the occurrence of Dupuytren's contracture, whereas agents preventing free radical release may prevent its occurrence.

Avoidance of Undesirable Side Effects

In the 1990s reports began to emerge suggesting that significantly fewer side effects were observed in humans when drugs displayed characteristics of a partial agonist and a partial antagonist rather than a full agonist (Im et al., 1995; Jacobsen et al., 1996a, Jacobsen et al., 1996b; Jacobsen et al., 1999). Partial agonists/antagonists often display inverted–U-shaped dose-response relationships. The decreased risk of developing undesirable side effects from partial agonists/antagonists was hypothesized to result from a lower capacity to induce responses at nontarget tissues. The inverted–U-shaped dose-response relationship of the partial agonist/antagonist also suggested a potentially broader therapeutic zone for drug optimization. This theoretical framework has been used by pharmaceutical companies in the search for synthetic agents with partial agonist/antagonist characteristics that display hormetic-like inverted–U-shaped dose-response relationships.

The fact that pharmaceutical companies were searching for hormetic-like inverted U-shaped dose-response relationships to minimize the undesirable side effects of drugs suggested that this solution may have been already "discovered" in the process of natural selection. Indeed, many endogenous agonists are partial agonists/antagonists with dose-response characteristics of the inverted–U-shaped type. This suggests that another reason for the selection of the hormetic dose-response relationship and its widespread generalization is to minimize the occurrence of undesirable side effects from endogenous agonists.

Environmental Risk Assessment

The field of risk assessment developed the concept of safety factors based on the threshold dose-response model in the 1920s. For example, in 1925 the threshold dose-response model was first employed for the protection of workers from exposure to radiation (Mutscheller, 1925). In the case of radiation, fear of cancer led in the mid 1950s to the replacement of the threshold dose-response model with a model that was linear at low doses (NCRPM, 1954). The FDA adopted the use of the threshold dose-response model along with safety factors (now called uncertainty factors) by the mid 1950s (Lehman, 1954). The threshold dose-response model continued to dominate chemical toxicology and risk assessment for all toxic endpoints until the U.S. National Academy of Sciences Safe Drinking Water Committee in 1977 recommended to the U.S. Environmental Protection Agency that they follow the lead of the radiation community and accept a linear-at-low-dose modeling approach for estimating risks from carcinogens while retaining the threshold dose-response model for noncarcinogens (NAS, 1977).

The use of linear-at-low-dose modeling for carcinogen risk assessment has been controversial principally because it can be very conservative in its estimation of cancer risks. As a result of this approach, health regulations for workers and community health have been enormously expensive as governmental agencies in the United States have tried to achieve *de minimus* risk goals such as no more than 1 excess cancer per 1 million people per 70-year lifetime. A major problem with this approach is that such model-based risk predictions cannot be practically validated experimentally or epidemiologically. In an attempt to determine the shape of the dose-response relationship of a genotoxic carcinogen (2-acetylaminofluorene [2-AAF]) in the low-dose zone, the FDA conducted what is now called the megamouse study in which 24,000 animals were used. In the end, the level of detectable risk was very insensitive, only at the level of 1 cancer per 100 people. This striking insensitivity (i.e., practical failure) of experimental systems to validate model estimates of cancer risks at and below 1 in 1,000 created a serious potential credibility problem for regulatory agencies. It indicated that risks of 1 in 1 million that agencies often use when communicating with the public about acceptable risks is a theoretical mathematical construct that cannot be practically tested, confirmed, or rejected. Such estimates are based on an unverifiable "belief system" based on which biostatistical model is more likely to be correct in its low-dose predictions based on current understandings of biological plausibility, arguments that also are not without their own substantial degree of uncertainty.

Due to the significance of the megamouse study and its potential impact on risk assessment practices in the United States and other countries, the U.S. Society of Toxicology (SOT) created a 14-member expert panel to evaluate this study's methods and data and devoted almost an entire issue of its journal, *Fundamental and Applied Toxicology*, to its analyses (Bruce et al., 1981). Of particular importance was that the SOT expert panel concluded that this carcinogen displayed unequivocal evidence of a hormetic dose-response relationship for bladder cancer (Fig. 7). The expert panel clearly stated that not only was there a threshold for the cancerous effect, but also that below the threshold the risk notably declined below background, as predicted by the hormetic dose-response model. This J-shaped dose response was consistently seen in each of the six rooms in which the large number of animals was housed, which thereby provided a type of quasi–built-in replication. Thus, in the largest chronic rodent bioassay for cancer, the hormetic dose-response relationship was observed. This study not only supported the finding of a hormetic dose-response relationship, but also revealed that linear-at-low-dose modeling predictions could not be tested at risk levels below 1 in 100.

Similar findings of J-shaped dose responses for numerous carcinogenic and mutagenic agents also have been reported. For example, consider the case of

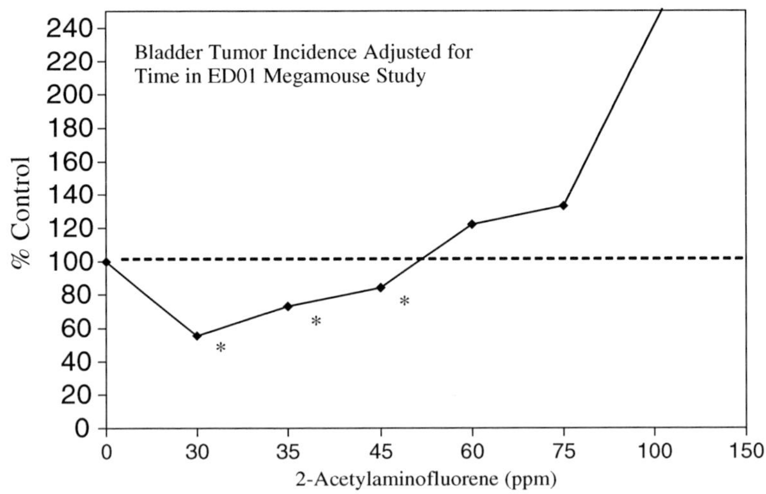

Fig. 7 Bladder tumor incidence adjusted for time in the ED01 megamouse study (Bruce et al., 1981)

dichlorodiphenyltrichloroethane (DDT), the banned pesticide, which the EPA regulates as a liver carcinogen based on linear-at-low-dose modeling. Considerable research indicates that DDT is a liver carcinogen in rodents, but only at high doses. As the dose is progressively decreased, the risk of liver cancer decreases, with the dose-response relationship becoming hormetic (Fig. 8)

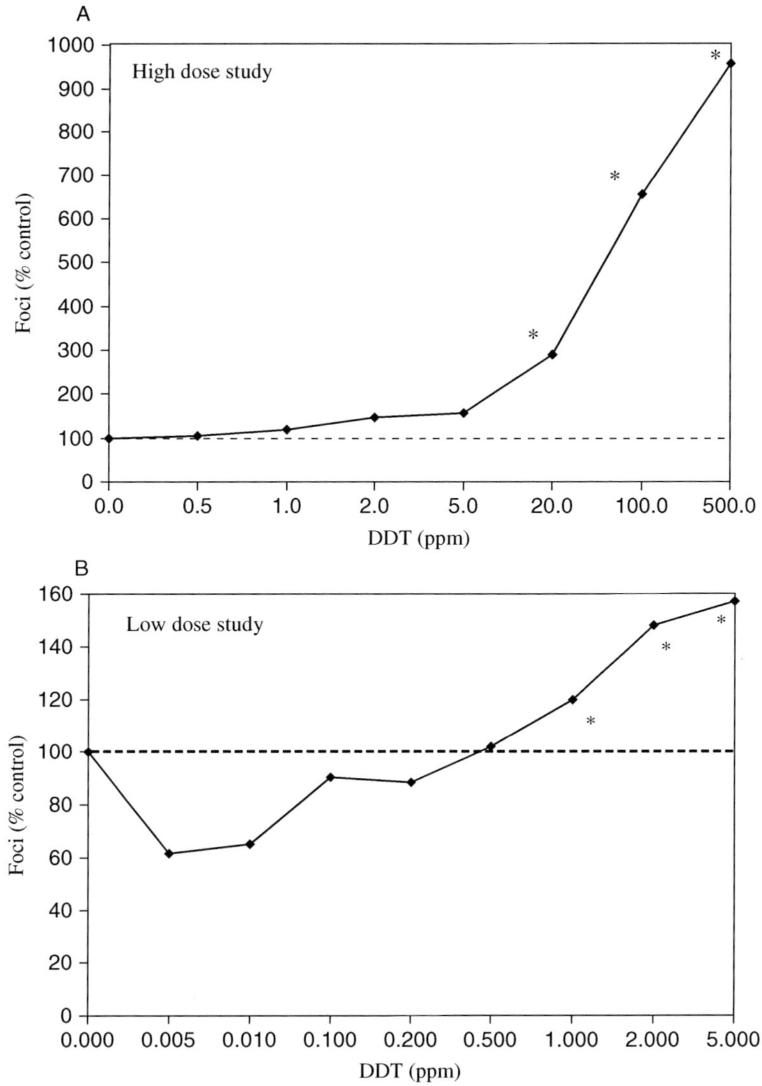

Fig. 8 Effect of dichlorodiphenyltrichloroethane (DDT) on the number of placental glutathione S-transferase (GST-P)–positive foci in F344 rat livers in two bioassays assessing different but slightly overlapping doses of carcinogen. Note: As the dose decreases, the J-shaped dose-response relationship becomes evident. Also note the difference in scale between the two graphs. (Sukata et al., 2002.)

(Fukushima et al., 2005; Sukata et al., 2002). Follow-up mechanistic research has clarified the underlying mechanistic reasons for the cancerous responses at high doses and the chemoprotective effects at lower doses. As was the case with 2-AAF, the linear-at-low-dose prediction of DDT-induced hepatic foci was not validated in experimental studies, whereas the hormetic dose-response model was.

A question of considerable societal importance is whether regulatory agencies should continue to use the threshold and linear-at-low-dose models in environmental risk assessment practices, especially given that both are ineffective in predicting responses in the low-dose zone (Calabrese and Baldwin, 2001; Calabrese et al., 2006; Calabrese and Baldwin, 2003b). It is highly questionable public policy to use toxicological models for risk assessment purposes that fail to accurately predict responses in the low-dose zone. Alternatively, the hormetic dose-response model could be considered for use in risk assessment practices because it can be tested and has performed well in the same tests at which the threshold and linear models have failed (Calabrese, 2005d). The hormetic dose-response model offers a range of other advantages for use in risk assessment, as indicated in Table 3.

Although the clinical challenges of trying to determine optimal performance doses can be difficult, they are fundamentally different from those with chemical risk assessment in the environmental regulatory arena. In the case of risk assessments for noncarcinogens, the standard practice has been to use multiple independent uncertainty factors. The procedures use a factor of 10 to account for the uncertainty of extrapolating from the average response of the animal model to the average human. Another uncertainty factor of 10 accounts for human interindividual variation ranging from the average human to those at increased risk. The use of these two independent uncertainty factors provides protection for the normal and high-risk segments of the human population. The hormesis concept refines this traditional dose-response/risk management methodology. Calabrese and Baldwin (Calabrese and Baldwin, 2002) found that normal and high-risk subgroups display hormetic dose-response relationships, with the high-risk group's dose-response relationship being shifted to the left due to their enhanced susceptibility. An optimal dose for the normal population may be in the toxic zone for the high-risk group, whereas the optimal dose for the high-risk group may be without effect for the normal population. If the goal were to optimize the health response in the overall population, there would be a conflict between the two interest groups, the normal and those at higher risk. Under the nonhormesis scenario, there is no such conflict because the process is simply to lower the dose below predicted risks, ignoring possible benefits. Yet the hormetic dose response model is the more biologically plausible situation. Thus, it is expected that in the future regulatory agencies will have to address the issue of health optimization as a policy option on the basis of the hormetic dose response.

Table 3 Hormesis: How It Affects Biology and Medicine

Impact on biological concepts
Performance – Low-dose stimulation is a measure of biological performance
Plasticity – The magnitude of the stimulatory response is an index of biological plasticity
Allometry – Hormesis responses conform to allometric analysis
Adaptation – Compensatory stimulation is a fundamental example of adaptation
Evolution – The hormetic response is highly conserved
Stress – Low-level stress optimizes system performance

Impact on toxicology/pharmacology concepts
Dose-response features – There is a well-defined magnitude of stimulation
Dose-time effect – The time component is needed to study compensatory response
Potency – The hormetic response is independent of potency
Hormetic synergy – Defines chemical interaction below the threshold
Mechanism strategy – There are multiple proximate mechanisms of hormesis
Study design – Modest low-dose stimulation supports unbalanced study designs
Study replication – Increased need to demonstrate reproducibility of findings

Environmental health risk assessment applications
Validation of predictions – Occur in observable zone; can be tested
Predict and quantify harm below threshold – No other model does this
Predict and quantify benefits below threshold – No other model does this
Biological model selection – Based on capacity to detect harm and benefit
Harmonization of cancer and noncancer risk assessment via hormesis – Occurs because dose-response characteristics of cancer and noncancer endpoints are similar with the hormetic model and independent of mechanism
Change risk goals – Optimizes population health, not only avoids harm
Cost/benefit – By including benefits hormesis, changes cost/benefit methods and strategies
Biostatistical modeling – Eliminates constraining models to pass through the origin
Uncertainty factor application methodology – Redesigned to optimize population health

Medical implications
Drug discovery and development – The hormesis effect defines potential value

(continued)

Table 3 (continued)

Therapeutic window – The width of hormetic stimulation for performance endpoints
Clinical trials – Need to incorporate U-shaped dose-response relationships in study design

Specific treatment areas
Aging – Low doses of stress extend longevity via hormetic mechanisms
Bone – Biphosphonates act hormetically, strengthening bones at low doses
Benign prostatic hyperplasia (BHP) – Various drugs induce BHP via hormetic dose response
Cancer – Numerous antitumor drugs cause proliferation of tumor cells at low doses
Cardiovascular disease – Low doses of statins act via hormesis on the vasculature
HIV treatment – Low doses of antiviral drugs cause proliferation of the virus
Memory – The action of most memory drugs follow U-shaped dose-response relationships
Neuroprotection – Numerous agents protect neurons hormetically
Ocular – Low doses of drugs can enhance the risk of retinal detachment
Pain – Many pain-relieving drugs often act biphasically via hormetic mechanisms
Prion diseases – Antiprion drugs can enhance disease at low doses
Sexual behavior – Most drugs enhancing male sexual performance act hormetically
Skin/hair – Minoxidil grows hair via hormetic processes
Antibiotics – Low doses of many antibiotics may enhance bacterial colony growth at low doses
Pulmonary hypertension – Low doses of the widely used anxiolytic drug Prozac increase the risk of pulmonary hypertension in the fetus by enhancing the proliferation of smooth muscle in the pulmonary arteries
Fibrotic conditions (e.g., Dupuytren's contracture of the hand) – Low doses of oxygen free radicals enhance the proliferation of human fibroblasts, leading to increased fibroblast density and pathological consequences

(continued)

Table 3 (continued)

Agriculture and related industrial applications
Growth rates – Low doses of stressor enhance growth in poultry
Animal grazing – Enhances plant productivity via hormetic processes
Allelochemistry – Chemicals released from plant roots act hormetically
Pesticide drift – Low doses of herbicides can enhance plant growth
Crop production – Hormetic mechanisms can increase growth
Reduced diseases – Plant diseases are reduced via hormetic mechanisms
Microbial metabolism – Hormetic mechanisms enhance greenhouse gas uptake in microorganisms
Bio-fuel cells – Hormetic mechanisms enhance hydrogen capture for bio-fuel cells in microorganisms
Drug production – Hormetic mechanisms enhance the synthesis of chemotherapeutic agents such as paclitaxel (Taxol)

Discussion

This chapter has argued that hormesis is the most fundamental dose-response relationship based on the outcomes of several large-scale head-to-head comparisons with its rival dose-response models (Calabrese and Baldwin, 2003a; Calabrese and Baldwin, 2001; Calabrese et al., 2006) along with its substantial reproducible occurrence in numerous publications across the spectrum of biomedical subdisciplines. This represents a striking change in perception, given that the hormetic dose-response model had, throughout the last century, been nearly completely marginalized and dismissed. This change is particularly noteworthy because the dose-response relationship is the central principle of disciplines such as toxicology and pharmacology. When professional scientific groups make an error in the central core principle of their discipline, this is a cause of considerable concern, especially in light of their role in guiding society's decisions concerning drug discovery, the safety evaluation of chemicals and drugs, and governmental risk assessment practices, which affect the derivation of environmental, occupational, and food safety standards.

The basis of this error in the understanding, selection, and use of dose-response models was complex, deriving in large part from the long-standing and bitter dispute between homeopathy and traditional medicine (Calabrese, 2005a). It was also due to inherent challenges in studying the hormetic dose response, with its need for stronger study designs, greater statistical power, and reproducibility of findings, all due to the fact that the hormetic stimulation is modest and in need of more rigorous evaluation and documentation. By denying that there are treatment-related effects below the toxicological and pharmacological thresholds, the field of toxicology developed an incorrect understanding of the nature of the dose-response relationship.

This chapter has provided a new and improved concept of the dose-response relationship. That is, the most fundamental nature of the dose response has a toxic component that begins as doses exceed the toxic threshold and a performance stimulation component that begins immediately below the threshold. The dual nature of the dose-response relationship, with the low-dose hormetic stimulation representing biological performance, is a novel interpretation. The low-dose performance stimulation has unique characteristics, with its maximum being modest, usually only 30% to 60% greater than the control value. This performance feature of the dose response provides a quantitative estimate of biological plasticity throughout biological systems at all levels of organization. Thus, the hormetic dose-response relationship is a basic, unifying and explanatory biological concept, in addition to being an important quantitative tool for the assessment of drugs, chemicals, and radiation.

The presence of performance and toxicity features of the dose response has important implications for numerous biological and biomedical disciplines (Table 3). For example, at high concentrations, antitumor drugs inhibit cell proliferation of tumor cell lines and other types of cells, whereas at lower concentrations these agents often display a stimulatory effect consistent with the quantitative features of the hormetic dose response. Similar responses also are commonly reported for antibiotics, antifungal agents, and antiviral drugs (Calabrese and Baldwin, 2003b). In a practical sense all medical treatments that are targeted for the toxicity zone of the dose-response relationship will eventually achieve a concentration within the hormetic or performance zone for a variable period of time due to pharmacokinetic factors. It is also important for those assessing the hazard potential of chemicals to define the entire dose-response continuum, which includes both the performance and toxicity dimensions. In general, the toxicological assessment of chemical agents only focuses on the above-threshold aspects of the dose response. Yet in all risk assessment practices, there are extrapolation procedures that estimate responses to doses far below the measured toxicological threshold without any information generated about the performance component of the dose response. In fact, the performance component of the dose response has been ignored in such cases under the incorrect assumption that it does not exist.

The overwhelming preponderance of evidence supports a conclusion that the long-revered threshold dose-response model fails to accurately predict responses in the below-threshold zone, that is, where people are routinely exposed. Continued reliance on this model to guide public regulatory judgments is no longer responsible public policy. Much evidence indicates that the hormetic dose-response model is highly generalizable and accurately predicts responses in the below-threshold zone, far outperforming the threshold and linear-at-low-dose models. Serious consideration should therefore be given to a major reevaluation of the continued use of current default dose-response models (i.e., threshold and linear) in the biological and health sciences, as well as in regulatory domains, especially in light of the failings of these models to predict low-dose effects, and their possible replacement with far more accurate and validatable models such as the hormetic dose-response model.

Acknowledgments This work was sponsored by the Air Force Office of Scientific Research, Air Force Material Command, USAF, under grant number FA9550-07-1-0248. The U.S. Government is authorized to reproduce and distribute reprints for governmental purposes notwithstanding any copyright notation hereon. The views and conclusions contained herein are those of the author and should not be interpreted as necessarily representing the official policies or endorsement, either expressed or implied, of the Air Force Office of Scientific Research or the U.S. Government.

References

Berliner H (1985) *A System of Scientific* Medicine. Philanthropic foundations in the Flexner era. New York: Tavistock Publications.

Bliss CI (1935) The calculation of the dosage-mortality curve. Ann Appl Biol 22: 134–167.

Bohme H (1986) Hugo Schulz (8/6/1853–7/13/1932) his life and work. Dissertation from the Medical Department of the Free University of Berlin. Translated from the German by Julia M. Ryan, University of Massachusetts, Amherst.

Bors J, Zimmer K (1970) Effects of low doses of x-rays on rooting and yield of carnation. Stimulation News 1 1: 16–21.

Boyera N, Galey I, Bernard BA (1997) Biphasic effects of minoxidil on the proliferation and differentiation of normal human keratinocytes. Skin Pharmacol 10: 206–220.

Branham SE (1929) The effects of certain chemical compounds upon the course of gas production by baker's yeast. J Bacteriol 18: 247–264.
Bruce RD, Carlton WW, Ferber KH, Hughes DH, Quast JF, Salsburg DS, Smith JM (Members of the Society of Toxicology ED_{01} Task Force) (1981) Re-examination of the ED_{01} study—adjusting for time on study. Fund Appl Toxicol 1: 67–80.
Calabrese EJ (2008a) Pharmacological enhancement of neuronal survival. Crit Rev Toxicol 38: 349–389.
Calabrese EJ (2008b) Enhancing and regulating neurite outgrowth. Crit Rev Toxicol 38: 391–418.
Calabrese EJ (2008c) Alzheimer's disease drugs: an application of the hormetic dose response model. Crit Rev Toxicol 38: 419–452.
Calabrese EJ (2008d) Pain and U-shaped dose responses: occurrence, mechanisms and clinical implications. Crit Rev Toxicol 38: 579–590.
Calabrese EJ (2008e) Stress biology and hormesis: the Yerkes-Dodson law in psychology: a special case of the hormesis dose-response. Crit Rev Toxicol 38: 453–462.
Calabrese EJ (2008f) An assessment of anxiolytic drug screening tests: hormetic dose responses predominate. Crit Rev Toxicol 38: 489–542.
Calabrese EJ (2008 g) Modulation of the epileptic seizure threshold: implications of biphasic dose responses. Crit Rev Toxicol 38: 543–556.
Calabrese EJ (2008h) Drug therapies for stroke and traumatic brain injury often display U-shaped dose responses: occurrence, mechanisms and clinical implications. Crit Rev Toxicol 38: 557–577.
Calabrese EJ (2008i) Converging concepts: adaptive response, preconditioning, and the Yerkes-Dodson law are manifestations of hormesis. Ageing Res Rev 7: 8–20.
Calabrese EJ (2008j) Hormesis and mixtures. Toxicol Appl Pharmacol 229: 262–263.
Calabrese EJ (2007) Threshold-dose-response model—RIP: 1911 to 2006. BioEssays 29: 686–688.
Calabrese EJ (2005a) Historical blunders: how toxicology got the dose-response relationship half right. Cell Mol Biol 51: 643–654.
Calabrese EJ (2005b) Hormetic dose-response relationships in immunology: occurrence, quantitative features of the dose response, mechanistic foundations, and clinical implications. Crit Rev Toxicol 35: 89–295.
Calabrese EJ (2005c) Cancer biology and hormesis: human tumor cell lines commonly display hormetic (biphasic) dose responses. Crit Rev Toxicol 35, 463–582.
Calabrese EJ (2005d) Toxicological awakenings: the rebirth of hormesis as a central pillar of toxicology. Toxicol Appl Pharmacol 204: 1–8.
Calabrese EJ (2003a) Chemotherapeutics and hormesis. Crit Rev Toxicol 33: 305–354.
Calabrese EJ (2003b) Inorganics and hormesis. Crit Rev Toxicol 33: 215–304.
Calabrese EJ (2001a) Apoptosis: biphasic dose responses. Crit Rev Toxicol 31: 607–614.
Calabrese EJ (2001b) Cell migration/chemotaxis: biphasic dose responses. Crit Rev Toxicol 31: 524–615.
Calabrese EJ (2001c) Dopamine: biphasic dose responses. Crit Rev Toxicol 31: 563–584.
Calabrese EJ (2001d) Nitric oxide: biphasic dose responses. Crit Rev Toxicol 31: 489–502.
Calabrese EJ (2001e) Estrogen and related compounds: biphasic dose responses. Crit Rev Toxicol 31: 503–516.
Calabrese EJ (2001f) Androgens: biphasic dose responses. Crit Rev Toxicol 31: 517–522.
Calabrese EJ (2001 g) 5-Hydroxytryptamine (serotonin): biphasic dose responses. Crit Rev Toxicol 31: 553–562.
Calabrese EJ (2001 h) Opiates: biphasic dose responses. Crit Rev Toxicol 31: 585–604.
Calabrese EJ (2001i) Adrenergic receptors: biphasic dose responses. Crit Rev Toxicol 31: 523–538.
Calabrese EJ (1991) Multiple chemical interactions. Chelsea, MI: Lewis Publishers.
Calabrese EJ (1985) Uncertainty factors and interindividual variation. Regul Toxicol Pharmacol 5: 190–196.
Calabrese EJ, Baldwin LA (2003a) The hormetic dose response model is more common than the threshold model in toxicology. Toxicol Sci 71: 246–250.

Calabrese EJ, Baldwin LA (2003b) Chemotherapeutics and hormesis. Crit Rev Toxicol 33: 305–353.

Calabrese EJ, Baldwin LA (2002) Hormesis and high risk groups. Regul Toxicol Pharmacol 35: 414–428.

Calabrese EJ, Baldwin LA (2001) The frequency of U-shaped dose-responses in the toxicological literature. Toxicol Sci 62: 330–338.

Calabrese EJ, Baldwin LA (2000a) Chemical hormesis: its historical foundations as a biological hypothesis. Hum Exp Toxicol 19: 2–31.

Calabrese EJ, Baldwin LA (2000b) The marginalization of hormesis. Hum Exp Toxicol 19: 32–40

Calabrese EJ, Baldwin LA (2000c) Radiation hormesis: its historical foundations as a biological hypothesis Hum Exp Toxicol 19: 41–75.

Calabrese EJ, Baldwin LA (2000d) Radiation hormesis: the demise of a legitimate hypothesis. Hum Exp Toxicol 19: 76–84.

Calabrese EJ, Baldwin LA (2000e) Tales of two similar hypotheses: the rise and fall of chemical and radiation hormesis. Hum Exp Toxicol 19: 85–97.

Calabrese EJ, Baldwin LA (1997) The dose determines the stimulation (and poison): development of a chemical hormesis database. Int J Toxicol 16: 545–559.

Calabrese EJ, Blain R (2005) The occurrence of hormetic dose responses in the toxicological literature, the hormesis database: an overview. Toxicol Appl Pharmacol 202: 289–301.

Calabrese EJ, Bachmann KA, Bailer AJ, Bolger PM, Borak J, Cai L, Cedergreen N, Cherian MG, Chiueh CC, Clarkson TW, Cook RR, Diamon DM, Doolittle DJ, Dorato MA, Duke SO, Feinendegen L, Gardner DE, Hart RW, Hastings KL, Hayes AW, Hoffmann GR, Ives JA, Jaworowski Z, Johnson TE, Jonas WB, Kaminski NE, Keller JG, Klaunig JE, Knudsen TB, Kozumbo WJ, Lettieri T, Liu S-Z, Maisseu A, Maynard KI, Masoro EJ, McClellan RO, Mehendale HM, Mothersill C, Newlin DB, Nigg HN, Oehme FW, Phalen RF, Philbert MA, Rattan SIS, Riviere JE, Rodricks J, Sapolsky RM, Scott BR, Seymour C, Sinclair DA, Smith-Sonneborn J, Snow ET, Spear L, Stevenson DE, Thomas Y, Tubiana M, Williams GM, Mattson MP (2007) Biological stress response terminology: integrating the concepts of adaptive response and preconditioning stress within a hormetic dose-response framework. Toxicol Appl Pharmacol 222: 122–128.

Calabrese EJ, Staudenmayer JW, Stanek EJ, Hoffmann GR (2006) Hormesis outperforms threshold model in NCI anti-tumor drug screening data. Toxicol Sci 94: 368–378.

Calder WA (1996) Size, function, and life history Mineola, NY: Dover Publications.

Chueh S-C, Guh J-H, Chen J, Lai M-K, Teng C-M (2001) Dual effects of ouabain on the regulation of proliferation and apoptosis in human prostatic smooth muscle cells. J Urol 166: 347–353.

Cicero TJ, Badger TM (1977) Effects of alcohol on the hypothalamic-pituitary-gonadal axis in the male rat. J Pharmacol Exp Ther 201: 427–433.

Clark AJ (1937) Handbook of experimental pharmacology. Berlin: Verlag Julius Springer.

Clark AJ (1933) Mode of action of drugs on Cells. London: Arnold.

Clark AJ (1927) The historical aspects of quackery, Part 2. Br Med J 2: 960.

Cookson MR, Mead C, Austwick SM, Pentreath VW (1995) Use of the MTT assay for estimating toxicity in primary astrocyte and C6 glioma cell cultures. Toxicol in Vitro 9: 39–48.

Crump T (2003) Contemporary medicine as presented by its practitioners themselves, Leipzig, 1923:217–250. Hugo Schulz. Nonlinearity Biol Toxicol Med 1: 295–318

Davies, JMS, Lowry, CV, and Davies, KJA (1995) Transient adaptation to oxidative stressing yeast. Arch Biochem Biophys 317: 1–6.

Flexner A (1910) Medical education in the United States and Canada. A report to the Carnegie Foundation for the Advancement of Teaching. New York.

Flood JF, Smith GE, Cherkin A (1985) Memory enhancement: supra-additive effect of subcutaneous cholinergic drug combinations in mice. Psychopharmacology 86: 61–67.

Flood JF, Smith GE, Cherkin A (1983) Memory retention: potentiation of cholinergic drug combinations in mice. Neurobiol Aging 4: 37–43.

Foekens JA, Sieuwerts AM, Stuurman-Smeets EMJ, Dorssers LCJ, Berns EMJJ, Klijn JGM (1992) Pleiotropic actions of suramin on the proliferation of human breast cancer cells in vitro. Int J Cancer 51: 439–444.

Foley GE,, Winter WD (1949) Increased mortality following penicillin therapy of chick embryos infected with *Candida albicans* var *Stellatoidea*. J Infect Dis 85: 268–274.

Fornaro E, Li D, Pan J, Belik J (2007) Prenatal exposure to fluoxetine induces fetal pulmonary hypertension in the rat. Am J Respir Crit Care Med 176: 1035–1040.

Fukushima S, Kinoshita A, Puatanachokchai R, Kushida M, Wanibuchi H, Morimura K (2005) Hormesis and dose-response-mediated mechanisms in carcinogenesis: evidence for a threshold in carcinogenicity of non-genotoxic carcinogens. Carcinogenesis 26: 1835–1845.

Gaddum JH (1962) Pharmacologists of Edinburgh. Annu Rev Pharmacol Toxicol 2: 1–10.

Gaddum JH (1933) Methods of biological assay depending on the quantal response. London: His Majesty's Stationery Office.

Garrod LP (1951) The reactions of bacteria to chemotherapeutic agents. Br Med J 1: 205–210.

Giuliani N, Pedrazzoni M, Negri G, Passeri G, Impicciatore M, Girasole G (1998) Bisphosphonates stimulate formation of osteoblast precursors and mineralized nodules in murine and human bone marrow cultures in vitro and promote early osteoglastogenesis in young and aged mice in vivo. Bone 22: 455–461.

Goerig M, Agarwal K, Esch JS (2000) The versatile August Bier (1861–1949), father of spinal anesthesia. J Clin Anesth 12: 561–569.

Honar H, Riazi K, Homayoun H, Sadeghipour H, Rashidi N, Ebrahimkhani MR, Mirazi N, Dehpour AR (2004) Ultra-low dose naltrexone potentiates the anticonvulsant effect of low dose morphine on clonic seizures. Neuroscience 129: 733–742.

Hunter PE, Krithayakiern V (1971) Effect of gamma radiation upon life expectancy and reproduction in the house cricket, *Acheta domesticus* (orthopter: Gryllidae). Ann Entomol Soc Am 64: 119–123.

Im HK, Im WB, Pregenzer JF, Carter DB, Jacobsen EJ, Hamilton BJ (1995) Characterization of U-97775 as a $GABA_A$ receptor ligand of dual functionality in cloned rat $GABA_A$ receptor subtypes. Br J Pharmacol 115: 19–24.

Jacobsen EJ, TenBrink RE, Stelzer LS, Belonga KL, Carter DB, Im HK, Im WB, Sethy VH, Tang AH, VonVoigtlander PF, Petke, JD (1996a) High affinity partial agonist imidazol[1,5-a]quinoxaline amides, carbamate, and ureas at the γ-aminobutyric acid A/benzodiazepine receptor complex. J Med Chem 39: 158–175.

Jacobsen EJ, Stelzer LS, Belonga KL, Carter DB, Im WB, Sethy VH, Tang AH, VonVoigtlander PF, Petke JD (1996b) 3-Phenyl-substituted imidazo[1,5-a]quinoxalin-4-ones and imidazol[1,5-a]quinoxaline ureas that have high affinity at the $GABA_A$/benzodiazepine receptor complex. J Med Chem 39: 3820–3836.

Jacobsen EJ, Stelzer LS, TenBrink RE, Belonga KL, Carter DB, Im HK, Im WB, Sethy VH, Tang AH, VonVoigtlander PF, Petke JD, Zhong WZ, Mickelson JW (1999) Piperazine imidazo[1,5-a]quinoxaline ureas as high-affinity $GABA_A$ ligands of dual functionality. J Med Chem 42: 1123–1144.

Lee JB, Hayashi K, Hayashi T, Sankawa U, Maeda M (1999) Antiviral activities against HSV-1, HCMV, and HIV-1 of rhamnan sulfate from *Monostroma latissimum*. Planta Med 65: 439–441.

Lehman AJ (1954) Untitled. Assoc Food Drug Off 18: 33.

Liberman UA, Weiss SR, Broll J, Minne HW, Quan H, Bell NH, Rodriguezportales J, Downs RW, Dequeker J, Favus M, Seeman E, Recker RR, Capizzi T, Santora AC, Lombardi A, Shah RV, Hirsch LJ, Karpf DB (1995) Effect of oral alendronate on bone mineral density and the incidence of fractures in postmenopausal osteoporosis. N Engl J Med 333: 1437–1443.

McGaugh JL, Petrinovich LF (1965) Effects of drugs on learning and memory. Int Rev Neurobiol 8: 139–196.

Melchior CL, Ritzmann RF (1994) Dehydroepiandrosterone is an anxiolytic in mice on the plus maze. Pharmacol Biochem Behav 47: 427–441.

Meng G (1993) Effects of arsenic on DNA synthesis in human lymphocytes. Arch Environ Contam Toxicol 25: 525–528.
Merkel LA, Lappe RW, Rivera LM, Cox BF, Perrone MH (1992) Demonstration of vasorelaxant activity with an A_1-selective adenosine agonist in porcine coronary artery: involvement of potassium channels. J Pharmacol Exp Ther 260: 437–443.
Miller WS, Green CA, Kitchen H (1945) Biphasic action of penicillin and other sulphonamide similarity. Nature 155: 210–211.
Murrell GAC, Francis MJO, Bromley L (1990) Modulation of fibroblast proliferation by oxygen free radicals. Biochem J 265: 659–665.
Murry CE, Jennings RB, Reimer KA (1986) Preconditioning with ischemia: a delay of lethal cell injury in ischemic myocardium. Circulation 75: 534–537.
Mutscheller A (1925) Physical standards of protection against Roentgen ray dangers. Am J Roentgenol 13: 65–69.
National Academy of Sciences (NAS) (1977) Drinking water and health. Washington, DC: National Academy of Sciences.
National Council on Radiation Protection and Measurements (NCRPM) (1954) Permissible dose from external sources of ionizing radiation. Washington, DC: U.S. Department of Commerce.
Nyberg K, Ekblad M, Bergstrom T, Freeman C, Parish CR, Ferro V, Trybala E (2004) The low molecular weight heparin sulfate-mimetic, PI-88, inhibits cell-to-cell spread of herpes simplex virus. Antiviral Res 63: 15–24.
Paalzow GHM, Paalzow LK (1985) Promethazine both facilitates and inhibits nociception in rats: effect of the testing procedure. Psychopharmacology 85: 31–36.
Paalzow GHM, Paalzow LK (1983a) Opposing effects of apomorphine on pain in rats. Evaluation of the dose-response curve. Eur J Pharmacol 88: 27–35.
Paalzow GHM, Paalzow LK (1983b) Yohimbine both increases and decreases nociceptive thresholds in rats evaluation of the dose-response relationship. Naunyn-Schmied Arch Pharmacol 322: 193–197.
Parkhurst BR, Bradshaw AS, Forte JL, Wright GP (1981) The chronic toxicity to *Daphnia magna* of acridine, a representative azarene present in synthetic fossil fuel products and wastewater. Environ Pollut (Series A) 24: 21–30.
Rai UN, Gupta M, Tripathi RD, Chandra P (1998) Cadmium regulated nitrate reductase activity in *Hydrilla verticillata* (1f) Royle. Water Soil Air Pollut 106: 171–177.
Randall WA, Price CW, Welch H (1947) Demonstration of hormesis (increase in fatality rate) by penicillin. Am J Public Health 37: 421–425.
Rossini M, Gatti D, Zamberlan N, Braga V, Dorizzi R, Adami S (1994) Long-term effects of a treatment course with oral alendronate of postmenopausal osteoporosis. J Bone Miner Res 9: 1833–1837.
Sandifer RD, Hopkin SP (1997) Effects of temperature on the relative toxicities of Cd, Cu, Pb, and Zn to *Folsomia candida* (collembolan). Ecotoxicol Environ Safety 37: 125–130
Schulz H (1888) Uber Hefegifte. Pfluger's Archiv Gesamte Physiol 42: 517–541.
Schulz H (1885) About the treatment of cholera nostras with veratrine [in German].Deutsch Arzneipflanzen Forsch Wochenzeit 11: 99–100. [Translated by Julia M. Ryan, University of Massachusetts, Amherst.]
Southam CM, Ehrlich J (1943) Effects of extract of Western red-cedar heartwood on certain wood-decaying fungi in culture. Phytopathology 33: 517–524.
Starke K (1998) A history of *Naunyn-Schmiedeberg's Archives of Pharmacology*. Naunyn-Schmied Arch Pharmacol 358: 1–109.
Sukata T, Uwagawa S, Ozaki K, Ogawa M, Nishikawa T, Iwai S, Kinoshita A, Wanibuchi H, Imaoka S, Funae Y, Okuno Y, Fukushima S (2002) Detailed low-dose study of 1,1-bis(*p*-chlorophenyl)-2,2,2-trichloroethane carcinogenesis suggests the possibility of a hormetic effect. Int J Cancer 99: 112–118.
Ungar J, Muggleton P (1946) The effect of penicillin on the growth of human type M tuberculosis. J Pathol Bacteriol 58: 501–504.

U.S. Environmental Protection Agency (U.S. EPA) (1986) Guidelines for the health risk assessment of chemical mixtures. Fed Regist 57: 34014–34025.

Verney EB, Barcroft J (1941) AJ Clark (Obituary). Obituary Notices Fellows. Roy Soc 3: 969.

Vieira VLP , Rocha JBT, Schetinger MRC , Morsch VM , Rodrigues SR , Tuerlinckz SM , Bohrer D, do Nascimento PC (2000) Effect of aluminum on δ-aminolevulinic acid dehydratase from mouse blood. Toxicol Lett 117: 45–52.

Welch H, Price CW, Randall WA (1946) Increase in fatality rate of *E. typhosa* for white mice by streptomycin. J Am Pharmaceut Assoc 35: 155–158.

Wiedman SJ, Appleby AP (1972) Plant growth stimulation by sublethal concentrations of herbicides. Weed Res 12: 65–74.

Wise LE, Lichtman AH (2007) The uncompetitive N-methyl-D-aspartate (NMDA) receptor antagonist memantine prolongs spatial memory in a rat delayed radial-arm maze memory task. Eur J Pharmacol 575: 98–102.

Wu W-C, Kao Y-H, Hu D-N (2002) A comparative study of effects of antiproliferative drugs on human retinal pigment epithelial cells in vitro. J Ocular Pharmacol Ther 18: 251–264.

The Fundamental Role of Hormesis in Evolution

Mark P. Mattson

Abstract Hormesis can be considered a major mechanism underlying Darwin's and Wallace's theory of evolution by natural selection. The ability of organisms to respond adaptively to low levels of exposure to environmental hazards in a manner that increases their resistance to more severe similar or different hazards is fundamental to the evolutionary process. The organisms that survive and reproduce are those best able to tolerate or avoid environmental hazards while competing successfully for limited energy (food) resources. Therefore many of the genes selected for their survival value encode proteins that protect cells against stress (heat-shock proteins, antioxidant enzymes, antiapoptotic proteins, etc.) or that mediate behavioral responses to environmental stressors (neurotransmitters, hormones, muscle cell growth factors, etc.). Examples of environmental conditions that can, at subtoxic levels, activate hormetic responses and examples of the genes and cellular and molecular pathways that mediate such adaptive stress responses are provided to illustrate how hormesis mediates natural selection.

Keywords Competition · Darwin · Ecology · Natural selection · Selenium · Stress resistance · Survival

Introduction

Life on Earth began in a hostile environment of limited (organic) resources and exposure to radiation, toxic metals, and other hazards (Williams, 2007). Today organisms also face many challenges that vary greatly, depending on the species

M.P. Mattson (✉)
Laboratory of Neurosciences, National Institutes of Health, National Institute on Aging, Baltimore, MD 21224, USA
e-mail: mattsonm@grc.nia.nih.gov

M.P. Mattson, E.J. Calabrese, *Hormesis*, DOI 10.1007/978-1-60761-495-1_3,

and its location and population density. The survival of the individual and its ability to pass genes on to the next generation depends on phenotypic traits that allow it to avoid or resist environmental stressors and compete successfully for limited resources (food, shelter, mates, etc.). Simple organisms such as bacteria and protozoa may die from exposures to toxic metals, dehydration, and excessively high or low temperatures. Evolutionary adaptations that guard against such hazards include toxin-impermeable membranes and cell motility mechanisms. In humans, the most highly evolved species, the major causes of death prior to reproductive age are infectious diseases, starvation, accidents, and homicide. Sophisticated innate and humoral immune systems, agricultural methods, seat belts, and police forces are examples of evolutionary adaptations that improve the chances of survival and reproduction in humans. In this chapter I describe evidence and a rationale for a hormesis-centric view of evolution.

The evolution of organic molecules, cells, multicellular organisms, and populations of organisms is characterized by an increase in complexity. The purposes of these complex biological systems are, in large part, to protect the cells and organisms against environmental stressors. Hormesis is any process in which exposure of a cell or organism to a sublethal dose of a stressor (chemical, thermal, energetic, psychological, etc.) activates adaptive stress response pathways that protect the cell/organism against more severe stresses of the same or different type (Fig. 1). In some situations hormesis pathways may be activated but ultimately fail to protect the cell or organism because of the severity and/or duration of the stress. In other situations (e.g., aging and disease states) the hormetic signaling pathways may be compromised. There are innumerable hormetic mechanisms that have evolved. For example, the development of lipid membranes with ion channels and pumps allowed cells to tightly control the intracellular ion concentrations and prevent toxic overloading with Na^+ and Ca^{2+} (Gotoh et al., 2007; Thomas and Rano, 2007). The evolution of proteins with specific binding sites for metals such as iron, copper,

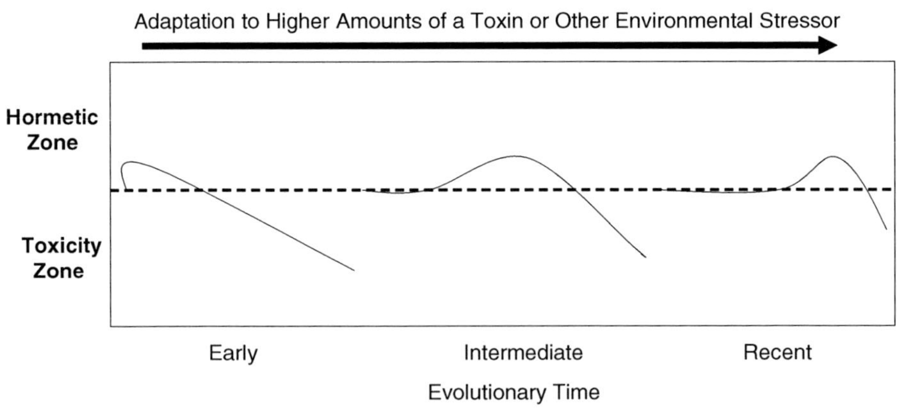

Fig. 1 The hormetic dose zone for exposures to toxins shifts to higher levels of exposures as the result of the evolution of novel mechanisms for toxin resistance, thus allowing organisms to occupy more stressful environments.

selenium, and zinc provided cells with a means of chelating these potentially toxic metals and also conferred new functional properties to the metal-binding proteins (copper/zinc superoxide dismutase, selenoproteins, ferritin, hemoglobin, and many others) (Crichton and Pierre, 2001).

The evolution of nervous systems allowed organisms to respond to environmental hazards by activating simple (reflexive withdrawal from a noxious agent) or complex (inventing seat belts and airbags to reduce the chances of injury and death in an automobile accident) behavioral responses. In mammals the "flight-or-fight" response includes the robust activation of neuroendocrine signaling pathways involving the brain, hypothalamus, pituitary gland, and adrenal gland, as well as the autonomic nervous system (Kopin, 1995). The immediate result of this adaptive stress response is the mobilization of energy reserves in the liver for use by the musculoskeletal and cardiovascular systems. Two key hormones that mediate the latter physiological changes are cortisol and epinephrine, which are produced by cells in the adrenal gland. These changes maximize the chance that the animal evades or withstands the challenge (predator, forest fire, etc.). However, sustained activation of the neuroendocrine stress response (as may occur under conditions of chronic psychosocial stress) can result in impaired hormesis and dysfunction and deterioration of tissues, resulting in pathological conditions such as cardiovascular disease, diabetes, osteoporosis, and psychiatric disorders (Chrousos, 2000). This example highlights the fact that both the magnitude of a stressor and its duration determine whether the organism is successful in responding to the stressor; a recovery period is often required for a hormetic response to be successful.

The Biphasic Dose Response and Evolution

Hormesis is a process in which there is a biphasic dose response to a natural or experimental perturbation of a cell or organism typified by a low-dose stimulatory or beneficial effect and a high-dose inhibitory or toxic effect (Calabrese et al., 2007; Mattson and Calabrese, 2008). The term hormesis is most widely used in the toxicology and biomedical fields, where investigators use it to describe biphasic dose responses of cells or organisms to toxins such as heavy metals, pesticides petrochemicals, and so on (Calabrese and Blain, 2005; Calabrese et al., 2007). Meta-analysis of data from research in the fields of toxicology, cancer biology, diet, neuroscience, drug development, and other areas have revealed the widespread existence of biphasic dose responses (Calabrese and Blain, 2005). The response of the cell or organism to the low dose of the toxin typically involves an adaptive compensatory process following an initial disruption in homeostasis. Thus, a short working definition of hormesis is "a process in which exposure to a low dose of a chemical agent or environmental factor that is damaging at higher doses induces an adaptive beneficial effect in the cell or organism." Several different terms are commonly used to describe specific types of hormetic responses, including "preconditioning" and "adaptive stress response" (Calabrese et al., 2007). The prevalence in the literature of hormetic dose responses to environmental toxins has been reviewed comprehensively (Calabrese and Blain, 2005), as have the implications of toxin-mediated hormesis for understanding carcinogenesis and its prevention (Calabrese, 2005).

The biphasic dose response has rarely been considered as an important aspect of evolution. Nevertheless, Parsons (2001) described hormesis in the context of ecology and evolution. Parsons proposed that "Fitness varies nonlinearly with environmental variables . . . with maximum fitness at intermediate levels between more stressful extremes," and that in the case of toxic agents, fitness is maximized at low concentrations. Parsons suggested that organisms inhabit environments for which they have evolved broad biological mechanisms to cope with the various stressors in that environment, so-called "hormetic zones." At first approximation it would seem reasonable to assume that organisms would survive best in environments where stress levels are low and so the minimum amount of energy is expended in resisting stressors. However, because energy (food) resources are limited, the energy expended in competition for those energy resources may be greater than the energy expended in counteracting stressors in a harsher but less populated area. For example, the plant prince's plume (*Stanleya pinnata*) accumulates high levels of the toxic element selenium, which protects it from caterpillar herbivory. However, an invasive species of diamondback moth (*Plutella xylostella*) has evolved a mechanism to withstand the selenium toxicity and thrives on prince's plume plants containing highly toxic selenium levels (Freeman et al., 2006), thus accessing a food source unavailable to other insects. Similarly, certain species of proteobacteria exhibit resistance to high levels of selenite, tellurite, and other rare-earth oxides and so can survive in soils where other species die, and so benefit from reduced competition for energy resources in those environments (Moore and Kaplan, 1992). Thus, different species of organisms have evolved to survive and reproduce in particular environments such that essentially all land and aquatic environments on Earth are now inhabited.

Plants spread slowly across the landscape through dispersion of seeds and growth of root shoots. By virtue of selection for genes that enhance hormetic pathways, plants can over time inhabit soils and climates with levels of stressors that would have killed their ancestors. Because they are not motile and therefore cannot escape stressors such as heat and cold, drought, or herbivores, a large portion of the genome of plants encodes proteins involved in protecting them against environmental extremes and consumption. For example, rice plants contain more genes than do humans, and many of the rice genes encode proteins that function in adaptive stress response pathways (Cooper et al., 2003). Microarray analyses of gene expression responses to various stressors have revealed conserved subcellular stress response pathways, as well as sets of genes that respond to certain types of stressor and not others (Hoffmann and Willi, 2008). Biphasic dose responses to a range of environmental factors have been documented in studies (Calabrese and Blain, 2008). The low-dose adaptive responses are likely mediated by evolutionarily conserved hormetic signaling pathways.

In my view the future of human evolution is unclear in part because advances in technology have led to a reduction in exposures to challenges that stimulate adaptive cellular stress responses. Such challenges include exercise, dietary energy restriction, and exposures to hot and cold temperatures. It is now clear that many of the major diseases that result in premature death are caused, in part, by a sedentary lifestyle combined with overeating. The chapter in this book entitled Couch Potato: The Antithesis of Hormesis elaborates on the downside of avoidance of hormetic challenges to the body and brain. However, at this point in human evolution (Table 1) the reduction in hormetic challenges appears not to be having a negative impact on reproduction, and one would therefore expect that selection for individuals with superior adaptive stress response mechanisms may not occur. Consequently, the postreproductive health and longevity of humans may actually decrease in the future as deleterious mutations accumulate (Parsons, 2003). The greatest challenges being faced by humans are largely psychological, and, accordingly, there has been an increase in the prevalence of psychiatric problems, including depression and anxiety and bipolar disorders (Kessler et al., 2007). Of interest, emerging evidence suggests that there is a shared neurochemical/neuroendocrine mechanism underlying the psychiatric disorders and poor energy metabolism (i.e., insulin resistance and diabetes). The evidence is as follows: (1) depression and anxiety disorders, as well as metabolic syndrome and diabetes, are associated with reduced serotonergic signaling and decreased levels of brain-derived neurotrophic factor (BDNF) in the brain (Krabbe et al., 2007); (2) exercise and dietary energy restriction increase BDNF levels in the brain and improve glucose regulation (Mattson et al., 2004a; Yamanaka et al., 2008); (3) exercise improves symptoms in patients with depression, anxiety, and bipolar disorders (Barbour et al., 2007); and (4) antidepressants that increase serotonin and BDNF signaling (serotonin reuptake inhibitors) also improve glucose regulation (McIntyre et al., 2006). Thus, the neurotransmitter serotonin and the neurotrophic factor BDNF can be considered as key mediators of hormetic responses to exercise and antidepressants. Although increased levels of serotonin and BDNF often have beneficial effects on neurons, excessive activation of serotonin and BDNF receptors can adversely affect the plasticity and survival of neurons (McDonald et al., 2002; Capela et al., 2007), consistent with biphasic dose response effects of these two mediators of hormesis.

Table 1 Examples of Toxic Substances, and the Adaptations That Cells and Organisms Have Evolved to Cope With or Utilize These Substances

Substance	Adaptation
O_2	Electron transport chain, antioxidant enzymes
CO_2	Respiratory exchange
CO	Guanylate cyclase, hemoproteins
NO	Guanylate cyclase
Fe^{2+}	Ferritin, transferrin
Cu^{+}	Ceruloplasmin
Ca^{2+}	Membranes, ion channels, transporters, binding proteins
H_2S	Sulfide dehydrogenase
UV radiation	Pigments

Cellular and Molecular Hormetic Mechanisms

To avoid extinction, organisms have developed complex mechanisms to cope with the environmental hazards they have encountered. Typically, such hormetic response pathways in cells involve proteins such as ion channels, kinases and deacetylases, and transcription factors that regulate the expression of genes that encode cytoprotective proteins (Mattson and Cheng, 2006). Examples of such pathways include receptors for the neurotransmitter glutamate in neurons that are coupled to calcium influx and activation of the transcription factors CREB and AP1 (Marini et al., 2008); receptors in muscle cells for acetylcholine that are sodium channels that when activated depolarize the plasma membrane, resulting in calcium influx (Booth, 1988); insulin receptors in liver cells coupled to the PI3 kinase–Akt kinase–FOXO transcription factor pathway (Matsumoto et al., 2006); increased production of reactive oxygen species (superoxide and hydrogen peroxide), resulting in the activation of the transcription factor Nrf-2 (Kang et al., 2005); and a reduction in cellular energy (ATP and NAD^+) levels, resulting in the activation of AMP kinase and inhibition of mTOR kinase (Martin and Hall, 2005). These pathways can be affected by behavioral responses involving the nervous system (exercise, neuroendocrine stress response activation, etc.) (McEwen, 2007), by ingestion or exposure to a noxious chemical (including chemicals in fruits, vegetables, and other plants) (Mattson and Cheng, 2006), or by reduced energy intake (Martin et al., 2006).

The genes induced by hormetic stressors include those encoding several categories of stress resistance proteins, including protein chaperones such as the heat-shock proteins, antioxidant enzymes such as superoxide dismutases and glutathione peroxidase, and growth factors such as insulin-like growth factors (IGFs) and brain-derived neurotrophic factor and BDNF (Mathers et al., 2004; Mattson et al., 2004 ; Young et al., 2004). Protein chaperones bind to other proteins, thus preserving their structure and protecting them against oxidative damage. For example, heat-shock protein (HSP-70) protects neurons against ischemic injury by stabilizing and enabling the function of antiapoptotic proteins such as Bcl-2 (Yenari et al., 2005). Other studies have shown that HSP-27 protects cells against oxidative stress by increasing glutathione levels and reducing levels of intracellular iron (Arrigo et al., 2005) (Fig. 2). Aerobic exercise results in the activation of mitogen-activated protein (MAP) kinases and the transcription factor NF-κB, which induces the expression of the antioxidant enzyme manganese superoxide dismutase (Kramer and Goodyear, 2007). NF-κB mediates hormetic responses to a variety of insults, including traumatic injury, infections, and oxidative stress (Mattson and Meffert, 2006). An evolutionarily conserved mechanism by which cells under stress warn adjacent cells of worsening conditions involves the production of one or more growth factors by the stressed cells. The growth factors typically activate receptor tyrosine kinases coupled to the PI3 kinase–Akt or MAP kinase pathways, resulting in the activation of transcription factors that induce the expression of cytoprotective proteins. For example, during a stroke, brain cells produce fibroblast growth factor (FGF), IGF-1, and BDNF, all of which act on neurons so as to increase the resistance of the cells to the metabolic and oxidative stress associated with

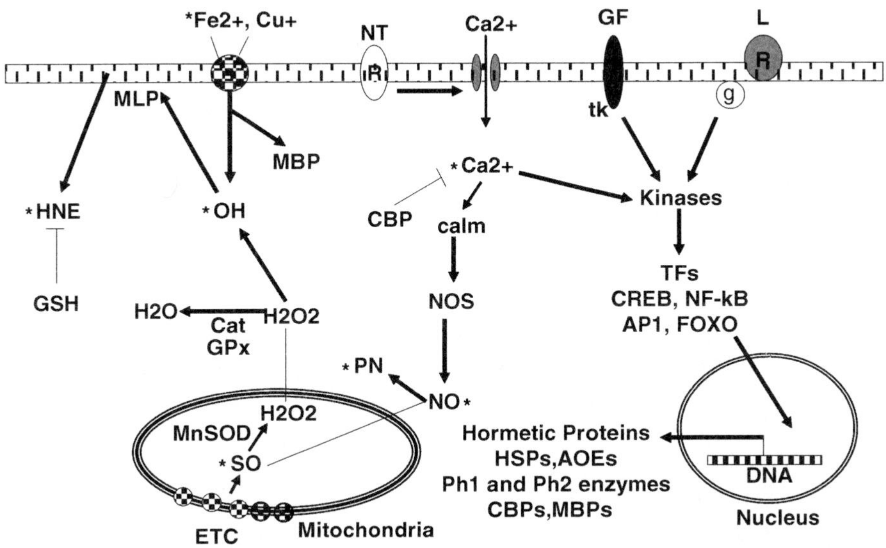

Fig. 2 Cells are continuously exposed to endogenous toxins that also serve important physiological roles, and so the cells have evolved mechanisms to limit the level of the toxins within a hormetic range of concentrations. Examples of toxic substances (marked with an asterisk) include Fe^{2+}, Cu^{+}, Ca^{2+}, superoxide (SO), hydroxyl radical (OH), nitric oxide (NO), peroxynitrite (PN), and 4-hydroxynonenal (HNE). Hormetic mechanisms include proteins that bind the toxins, such as glutathione (GSH), metal-binding proteins (MBP), and calcium-binding proteins (CBP); antioxidant enzymes, such as manganese superoxide dismutase (MnSOD), catalase (Cat), and glutathione peroxidase (GPx); the activation of receptors for neurotransmitters (NT), growth factors (GF), and ligands (L) for G protein–coupled receptors, resulting in the activation of kinases and transcription factors (TFs), which induce the expression of cytoprotective proteins, including heat-shock proteins (HSPs), antioxidant enzymes (AOEs), phase 1 (Ph1) and phase 2 (Ph2) enzymes, CBPs, and MBPs. calm, calmodulin

the ischemia (Mattson et al., 2002, 2004; Yamanaka et al., 2008); without these growth factors many more nerve cells would die from the stroke (Mattson et al., 2000; Arumugam et al., 2009). Age-related neurodegenerative disorders such as Alzheimer's and Parkinson's diseases may, in part, result from impaired hormesis signaling mechanisms, including decreased production or activity of neurotrophic factors such as BDNF. The chapter in this book entitled Hormesis and Aging describe how a decrease in "homeodynamic space" (adaptive response capabilities) may be a fundamental aspect of aging.

Hormesis and Evolutionary Strategies: Diversification and Specialization

Two general ways in which genetic mutations/diversity foster the evolution of hormetic mechanisms are by expanding the functional repertoire of proteins and pathways (diversification) and by increasing the number of genes devoted to coping with a specific stressor (specialization). A plausible example of diversification is the evolution of "heat-shock" proteins that possess the ability to protect cells against not only thermal stress, but also oxidative, metabolic, and calcium-induced stress (Fehrenbach and Niess, 1999). As species evolved to inhabit different niches, it is likely that mutations were selected for that preserved the original beneficial properties of a heat-shock protein while extending the range of stressors to which it responds (Fig. 2). Toxin-metabolizing enzymes such as the cytochrome P450 s (CYPs) can be considered examples of a genetic diversification of a hormetic mechanism. CYPs have been most intensively studied in liver cells, where they are known to detoxify natural or man-made toxins/drugs (Campbell and Hayes, 1974; Schuetz, 2001). Indeed, in some cases dietary phytochemicals are metabolized by the same CYP(s) that metabolize commonly used drugs, a fact that is increasingly recognized as a potential confound in drug dosing because diet-derived chemicals may alter drug pharmacokinetics (Delgoda and Westlake, 2004). Although the liver is the major site of xenobiotic metabolism, CYPs are also expressed by cells in other organs. For example, brain cells express some CYPs and can metabolize a range of phytochemicals and drugs (Ravindranath et al., 1995). CYPs are typically induced by sublethal amounts of toxins and can therefore be considered as important mediators of hormetic responses. In this way, CYPs have played a major role in protecting cells and organisms against a range of toxic agents encountered during evolution and the dispersal of species as they populated new territories (Fig. 3).

It is clear that, in contrast to heat-shock proteins and CYPs, which deal with multiple stressors, some genes evolved to cope with a specific environmental stressor. For example, iron is highly toxic in its free (Fe^{2+}) form but is relatively harmless

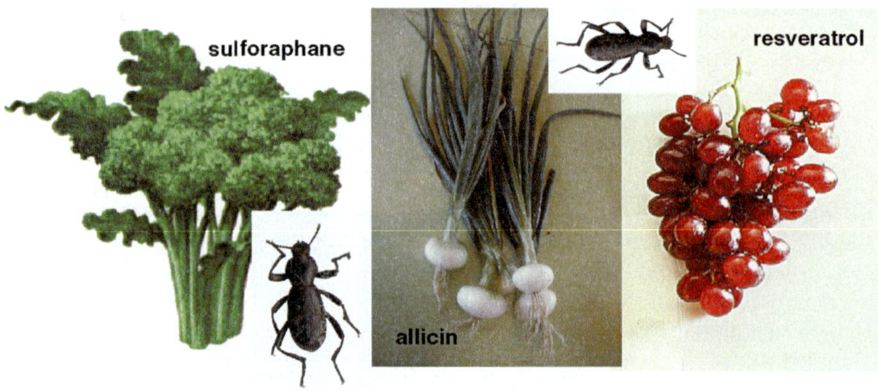

Fig. 3 Phytochemical hormesis. Many chemicals are produced by plants for the purpose of dissuading insects and microorganisms from eating their buds, leaves, fruit, and roots. However, when consumed in low amounts by animals and humans, some of these plant "toxins" can stimulate adaptive stress responses that protect against cancers, diabetes, cardiovascular disease, and neurological disorders. Examples of such phytochemicals and plants in which they are concentrated include sulforaphane (broccoli), allicin (onions and garlic), and resveratrol (red grapes and wine)

when bound to proteins. Because iron is ubiquitous in water sources and foodstuffs, organisms evolved several different proteins that specifically bind and transport or metabolize iron, including ferritin, transferrin, ferroportin, and ceruloplasmin (Mackenzie et al., 2008). Similarly, there are multiple genes that encode proteins that function to regulate other potentially toxic substances, including copper, zinc, and, most remarkably, calcium (calmodulin, calbindin, Ca^{2+}-ATPases, and many others) (Kawasaki et al., 1998; Gifford et al., 2007). The chemical properties and relative abundance of calcium and iron are presumably major reasons why so many genes evolved that encode proteins that bind to or otherwise control the movement and function of these ions. The consequent evolution of multiple physiological functions for calcium and iron has resulted in organisms, such as humans, that cannot survive without calcium or iron. With regard to hormesis, it is clear that cells and organisms possess very efficient mechanisms for maintaining levels of free iron and calcium at or below the hormetic concentration range. In the case of calcium it has been shown that moderate and transient increases in intracellular calcium levels mediate hormetic responses. An excellent example is given by neurons in the brain, in which calcium influx occurs in response to different mild stressors (e.g., activity associated with learning, body movements, or mild ischemia). Inside the neuron, calcium then activates kinases and transcription factors, resulting in the induction of genes that encode neuroprotective trophic factors, antioxidant enzymes, and cell repair proteins. However, excessive uncontrolled elevations of intracellular calcium overwhelm hormetic mechanisms and result in cell damage and death; such unleashing of toxic levels of calcium is believed to play a role in many different disorders, including Alzheimer's disease, myocardial infarction, and stroke (Bezprozvanny and Mattson, 2008; Przyklenk et al., 1999; Marini et al., 2007).

From the perspective of their roles in hormesis, the evolution of sensory organs provides examples of both specialization and diversification. The olfactory system includes receptors that are highly sensitive to low levels of many different substances, and those receptors communicate with neurons in the olfactory bulb that then relay the signal to multiple brain regions (Bargmann, 2006). In some cases an odor (pheromones from a predator, smoke from a fire) elicits an immediate flight-or-fight response, whereas other odors (volatile chemicals in foods or pheromones from the opposite sex of the same species) elicit feeding or mating behavioral programs. From a genetic/evolutionary perspective the mutations in genes encoding olfactory receptors (which are G protein–coupled receptors) increased the number of molecules (odors) that could be detected but also increased the specificity of the sensory organ (Niimura and Nei, 2006). Individuals with genes that increased the sensitivity of olfactory receptors to odors from predators, potential mates, and food sources would have a survival advantage. Similarly, mutations in genes encoding proteins involved in light perception resulted in the expansion of the range of wavelengths of light perceived and color vision, again increasing the ability of the organism to detect a "low level" of a potential hazard (predator at a distance) and so activate the appropriate hormetic behavioral response.

Conclusions and Future Directions

The fundamental role of adaptive stress responses in the process of evolution by natural selection was established by Charles Darwin (http://darwin-online.org.uk/contents.html) and has since been supported across all levels of life (molecular, cellular, organ systems, individuals, and populations). One might argue that nothing is gained by using the term hormesis to describe the importance of adaptive stress responses in evolutionary processes. However, as can be appreciated from examples presented in this chapter and throughout this book, the biphasic dose response that defines hormesis refines and informs the more general adaptive stress response theory of evolution.

Acknowledgments This work was supported by the Intramural Research Program of the National Institute on Aging, National Institutes of Health.

References

Alt FW (2007) Sirtuins and cellular stress responses. Ageing Res Rev this issue

Arrigo AP, Virot S, Chaufour S, Firdaus W, Kretz-Remy C, Diaz-Latoud C (2005) Hsp27 consolidates intracellular redox homeostasis by upholding glutathione in its reduced form and by decreasing iron intracellular levels. Antioxid Redox Signal 7: 414–422.

Arumugam TV, Phillips TM, Cheng A, Morrell CH, Mattson MP, Wan R (2009) Age and energy intake interact to modify cell stress pathways and stroke outcome. Ann Neurol. In press.

Barbour KA, Edenfield TM, Blumenthal JA (2007) Exercise as a treatment for depression and other psychiatric disorders: a review. J Cardiopulm Rehabil Prev 27: 359–367.

Bargmann CI (2006) Comparative chemosensation from receptors to ecology. Nature 444: 295–301.

Bezprozvanny I, Mattson MP (2008) Neuronal calcium mishandling and the pathogenesis of Alzheimer's disease. Trends Neurosci 31: 454–463.

Booth FW (1988) Perspectives on molecular and cellular exercise physiology. J Appl Physiol 65: 1461–1471.

Calabrese EJ (2005) Cancer biology and hormesis: human tumor cell lines commonly display hormetic (biphasic) dose responses. Crit Rev Toxicol 35: 463–582.

Calabrese EJ, Blain R (2005) The occurrence of hormetic dose responses in the toxicological literature, the hormesis database: an overview. Toxicol Appl Pharmacol 202: 289–301.

Calabrese EJ, Bachmann KA, Bailer AJ, Bolger PM, Borak J, Cai L, Cedergreen N, Cherian MG, Chiueh CC, Clarkson TW, Cook RR, Diamond DM, Doolittle DJ, Dorato MA, Duke SO, Feinendegen L, Gardner DE, Hart RW, Hastings KL, Hayes AW, Hoffmann GR, Ives JA, Jaworowski Z, Johnson TE, Jonas WB, Kaminski NE, Keller JG, Klaunig JE, Knudsen TB, Kozumbo WJ, Lettieri T, Liu SZ, Maisseu A, Maynard KI, Masoro EJ, McClellan RO, Mehendale HM, Mothersill C, Newlin DB, Nigg HN, Oehme FW, Phalen RF, Philbert MA, Rattan SI, Riviere JE, Rodricks J, Sapolsky RM, Scott BR, Seymour C, Sinclair DA, Smith-Sonneborn J, Snow ET, Spear L, Stevenson DE, Thomas Y, Tubiana M, Williams GM, Mattson MP (2007) Biological stress response terminology: integrating the concepts of adaptive response and preconditioning stress within a hormetic dose-response framework. Toxicol Appl Pharmacol 222: 122–128.

Calabrese EJ, Blain RB (2008). Hormesis and plant biology. Environ Pollut 87: 727–732.

Campbell TC, Hayes JR (1974) Role of nutrition in the drug-metabolizing enzyme system. Pharmacol Rev 26: 171–197.

Capela JP, Fernandes E, Remião F, Bastos ML, Meisel A, Carvalho F (2007) Ecstasy induces apoptosis via 5-HT(2a)-receptor stimulation in cortical neurons. Neurotoxicology 28: 868–875.

Chrousos GP (2000) The role of stress and the hypothalamic-pituitary-adrenal axis in the pathogenesis of the metabolic syndrome: neuro-endocrine and target tissue-related causes. Int J Obes Relat Metab Disord 24(Suppl 2): S505.

Cooper B, Clarke JD, Budworth P, Kreps J, Hutchison D, Park S, Guimil S, Dunn M, Luginbühl P, Ellero C, Goff SA, Glazebrook J (2003) A network of rice genes associated with stress response and seed development. Proc Natl Acad Sci USA 100: 4945–4950.

Crichton RR, Pierre JL (2001) Old iron, young copper: from Mars to Venus. Biometals 14: 99–112.

Delgoda R, Westlake AC (2004) Herbal interactions involving cytochrome p450 enzymes: a mini review. Toxicol Rev 23: 239–249.

Fehrenbach E, Niess AM (1999) Role of heat shock proteins in the exercise response. Exerc Immunol Rev 5: 57–77.

Freeman JL, Quinn CF, Marcus MA, Fakra S, Pilon-Smits EA (2006) Selenium-tolerant diamondback moth disarms hyperaccumulator plant defense. Curr Biol 16: 2181–2192.

Gifford JL, Walsh MP, Vogel HJ (2007) Structures and metal-ion-binding properties of the Ca2+-binding helix-loop-helix EF-hand motifs. Biochem J 405: 199–221.

Gotoh M, Sugawara A, Akiyoshi K, Matsumoto I, Ourisson G, Nakatani Y (2007) Possible molecular evolution of biomembranes: from single-chain to double-chain lipids. Chem Biodivers 4: 837–848.

Hoffmann AA, Willi Y (2008) Detecting genetic responses to environmental change. Nat Rev Genet 9: 421–432.

Kang KW, Lee SJ, Kim SG (2005) Molecular mechanism of nrf2 activation by oxidative stress. Antioxid Redox Signal 7:1664–1673.

Kawasaki H, Nakayama S, Kretsinger RH (1998) Classification and evolution of EF-hand proteins. Biometals 11: 277–295.

Kessler RC, Merikangas KR, Wang PS (2007) Prevalence, comorbidity, and service utilization for mood disorders in the united states at the beginning of the twenty-first century. Annu Rev Clin Psychol 3: 137–158.

Kopin IJ (1995) Definitions of stress and sympathetic neuronal responses. Ann N Y Acad Sci 771: 19–30.

Krabbe KS, Nielsen AR, Krogh-Madsen R, Plomgaard P, Rasmussen P, Erikstrup C, Fischer CP, Lindegaard B, Petersen AM, Taudorf S, Secher NH, Pilegaard H, Bruunsgaard H, Pedersen BK (2007) Brain-derived neurotrophic factor (BDNF) and type 2 diabetes. Diabetologia 50: 431–438.

Kramer HF, Goodyear LJ (2007) Exercise, MAPK, and NF-kappaB signaling in skeletal muscle. J Appl Physiol 103: 388–395.

Mackenzie EL, Iwasaki K, Tsuji Y (2008) Intracellular iron transport and storage: from molecular mechanisms to health implications. Antioxid Redox Signal. 10: 997–1030.

Marini AM, Jiang X, Wu X, Pan H, Guo Z, Mattson MP, Blondeau N, Novelli A, Lipsky RH (2007) Preconditioning and neurotrophins: a model for brain adaptation to seizures, ischemia and other stressful stimuli. Amino Acids 32: 299–304.

Marini AM, Jiang H, Pan H, Wu X, Lipsky RH (2008) Hormesis: a promising strategy to sustain endogenous neuronal survival pathways against neurodegenerative disorders. Ageing Res Rev 7: 21–33.

Martin DE, Hall MN (2005) The expanding TOR signaling network. Curr Opin Cell Biol 17: 158–166.

Matsumoto M, Han S, Kitamura T, Accili D (2006) Dual role of transcription factor FoxO1 in controlling hepatic insulin sensitivity and lipid metabolism. J Clin Invest 116: 2464–2472.

Mattson MP, Maudsley S, Martin B (2004a) BDNF and 5-HT: a dynamic duo in age-related neuronal plasticity and neurodegenerative disorders. Trends Neurosci 27: 589–594.

Martin B, Mattson MP, Maudsley S (2006) Caloric restriction and intermittent fasting: two potential diets for successful brain aging. Ageing Res Rev 5: 332–353.

Mathers J, Fraser JA, McMahon M, Saunders RD, Hayes JD, McLellan LI (2004) Antioxidant and cytoprotective responses to redox stress. Biochem Soc Symp 71: 157–176.
Mattson MP, Culmsee C, Yu ZF (2000) Apoptotic and antiapoptotic mechanisms in stroke. Cell Tissue Res 301: 173–187.
Mattson MP, Duan W, Chan SL, Cheng A, Haughey N, Gary DS, Guo Z, Lee J, Furukawa K (2002) Neuroprotective and neurorestorative signal transduction mechanisms in brain aging: modification by genes, diet and behavior. Neurobiol Aging 23: 695–705.
Mattson MP, Maudsley S, Martin B (2004a) BDNF and 5-HT: a dynamic duo in age-related neuronal plasticity and neurodegenerative disorders. Trends Neurosci 27: 589–594.
Mattson MP, Maudsley S, Martin B (2004b) A neural signaling triumvirate that influences ageing and age-related disease: insulin/IGF-1, BDNF and serotonin. Ageing Res Rev 3: 445–464.
Mattson MP, Cheng A (2006) Neurohormetic phytochemicals: low-dose toxins that induce adaptive neuronal stress responses. Trends Neurosci 29: 632–639.
Mattson MP, Meffert MK (2006) Roles for NF-kappaB in nerve cell survival, plasticity, and disease. Cell Death Differ 13: 852–860.
Mattson MP, Calabrese EJ (2008) Best in small doses. New Scientist 2008(August 9): 36–39.
Mcdonald JW, Stefovska VG, Liu XZ, Shin H, Liu S, Choi DW (2002) Neurotrophin potentiation of iron-induced spinal cord injury. Neuroscience 115: 931–939.
McEwen BS (2007) Physiology and neurobiology of stress and adaptation: central role of the brain. Physiol Rev 87: 873–904.
McIntyre RS, Soczynska JK, Konarski JZ, Kennedy SH (2006) The effect of antidepressants on glucose homeostasis and insulin sensitivity: synthesis and mechanisms. Expert Opin Drug Saf 5: 157–168.
Moore MD, Kaplan S (1992) Identification of intrinsic high-level resistance to rare-earth oxides and oxyanions in members of the class proteobacteria: characterization of tellurite, selenite, and rhodium sesquioxide reduction in *Rhodobacter sphaeroides*. J Bacteriol 174: 1505–1514.
Niimura Y, Nei M (2006) Evolutionary dynamics of olfactory and other chemosensory receptor genes in vertebrates. J Hum Genet 51: 505–517.
Parsons PA (2001) The hormetic zone: an ecological and evolutionary perspective based upon habitat characteristics and fitness selection. Q Rev Biol 76: 459–467.
Parsons PA (2003) From the stress theory of aging to energetic and evolutionary expectations for longevity. Biogerontology 4: 63–73.
Przyklenk K, Simkhovich BZ, Bauer B, Hata K, Zhao L, Elliott GT, Kloner RA (1999) Cellular mechanisms of infarct size reduction with ischemic preconditioning. Role of calcium?. Ann N Y Acad Sci 874: 192–210.
Ravindranath V, Bhamre S, Bhagwat SV, Anandatheerthavarada HK, Shankar SK, Tirumalai PS (1995) Xenobiotic metabolism in brain. Toxicol Lett 82–83: 633–638.
Schuetz EG (2001) Induction of cytochromes P450. Curr Drug Metab 2: 139–147.
Thomas JA, Rana FR (2007) The influence of environmental conditions, lipid composition, and phase behavior on the origin of cell membranes. Orig Life Evol Biosph 37: 267–285.
Williams RJ (2007) A chemical systems approach to evolution. Dalton Trans 14: 991–1001.
Yamanaka M, Itakura Y, Ono-Kishino M, Tsuchida A, Nakagawa T, Taiji M (2008) Intermittent administration of brain-derived neurotrophic factor (BDNF) ameliorates glucose metabolism and prevents pancreatic exhaustion in diabetic mice. J Biosci Bioeng 105: 395–402.
Yenari MA, Liu J, Zheng Z, Vexler ZS, Lee JE, Giffard RG (2005) Antiapoptotic and anti-inflammatory mechanisms of heat-shock protein protection. Ann N Y Acad Sci 1053: 74–83.
Young JC, Agashe VR, Siegers K, Hartl FU (2004) Pathways of chaperone-mediated protein folding in the cytosol. Nat Rev Mol Cell Biol 5: 781–791.

Transcriptional Mediators of Cellular Hormesis

Tae Gen Son, Roy G. Cutler, Mark P. Mattson, and Simonetta Camandola

Abstract Hormesis is the beneficial adaptive response of cells and organisms to acute subtoxic doses of certain types of environmental stressors (e.g., heat, oxidation, environmental toxins). Repetitive hormesis through routine exercise, calorie restriction, or ingestion of low levels of phytotoxins with the diet can stimulate cellular catabolic turnover of damaged molecules and increase protective mechanisms. The net result is an improved ability of the organism to better cope with noxious insults (i.e., preconditioning). Key to the benefits of hormesis are (1) the intensity of the stress/toxin, which needs to be enough to stimulate an effective response without causing permanent damage (i.e., subtoxic) and (2) the duration of the exposure, which needs to be limited (acute) to allow repair and recovery. Fundamental to the hormetic adaptive response is gene expression regulation. Although different stressors elicit unique signature responses, the comparison of prototypical hormetic inducers has highlighted the role played by a few transcription factor families. The periodic pulsatile activation of Nrf2, NF-κB, HSF, and FOXO has been found to be essential to obtaining the beneficial effects of various hormetic stimuli in different biological models. This chapter discusses molecular mechanisms and gene targets for these transcription factor families in the hormetic adaptive context.

Keywords Transcription factors · Hormesis · Exercise · Calorie restriction · Phytochemicals · HSF · Nrf2 · FOXO · NF-κB

Introduction

Exposing organisms and cells to brief periods of mild stress renders them more resistant to the potential challenges of a subsequent, even greater stress. This phenomenon, known as hormesis, depends on the ability of cells and organism to upregulate their stress response–induced gene expression and the related pathways

T.G. Son (✉)
Laboratory of Neurosciences, National Institutes of Health, National Institute on Aging, Baltimore, MD 21224, USA
e-mail: sont2@grc.nia.nih.gov

M.P. Mattson, E.J. Calabrese, *Hormesis*, DOI 10.1007/978-1-60761-495-1_4,

of maintenance and repair. The validity of the hormetic concept has been provided by experiments with various biological systems and by using various chemical, physical, and biological stressors. Paradigms of hormetic stressors are exercise, calorie restriction, temperature shock, irradiation, and pro-oxidants (Minois, 2000; Rattan, 2004). Exercise represents an oxidative and metabolic stressor for the skeletal muscle and cardiovascular systems by generating free radicals, acids, and other damaging intermediates (McArdle et al., 2002). However, or rather because of it, moderate daily exercise clearly improves the quality of life and prevents disease (Rattan, 2004). In a similar way, organisms perceive nutrient limitation as a metabolic stress, yet calorie restriction has been shown to extend the lifespan and decrease morbidity in organisms as evolutionarily distant as *Caenorhabditis elegans* and primates (Kirkwood and Shanley, 2005; Bishop and Guarente, 2007a).

The ability of cells and organisms to adapt to short- and long-term environmental changes by modifying their gene expression is achieved through several transcriptional and posttranscriptional mechanisms, including chromatin remodeling and transcription; cell signaling; mRNA splicing, polyadenylation, and localization; and mechanisms of protein localization, modification, and degradation. Because transcription is a pivotal event in most adaptive stress responses, much emphasis and effort have been placed on understanding transcriptional regulation mechanisms. Different stressors elicit unique expression patterns according to the nature, magnitude, and duration of the insult. As the signature mechanisms responsible for their effects have been unraveled, it has become increasingly clear that most, if not all, stress conditions share a conserved core response. Categories of stress-induced genes include those encoding heat-shock proteins and antioxidant enzymes and proteins involved in energy metabolism. Genes for which expression is repressed by stress include those involved in growth-related functions, thus reflecting a redirection of resources toward stress protection.

In this chapter we review the basic concept of transcriptional regulation and summarize recent knowledge on the transcription factor families that are emerging as key players in the adaptive hormetic response.

Nature of Transcriptional Regulation

Eukaryotic DNA resides in the nucleus in 23 pairs of chromosomes containing 6×10^9 base pairs and interacts with thousands of specific DNA-binding transcription factors. The basic chromatin structure is the nucleosome unit. Each nucleosome consists of a core histone octamer formed by a central heterotetramer of histones H3 and H4 sandwiched between a pair of H2A and H2B heterodimers. Each histone octamer has approximately 147 base pairs of supercoiled DNA wrapped around it, and the nucleosomes are separated by variable, species-specific DNA linkers of 28 to 43 base pairs. The chromatin fiber is folded into a more condensed 30-nm structure by a single molecule of histone H1 bridging together nucleosomal units by binding at the beginning, center, and end of the nucleosomal DNA. Multiple orders of coiling then give rise to the compact 100- to 400-nm "coiled-coil" structure of interphase chromatin called heterochromatin. Both the coiled-coil and nucleosomal structures are stable at the physiological ionic strength and prevent unwanted gene transcription. The transcriptional activation of gene expression is achieved through decondensation and chromatin remodeling. The less compact structure of the transcriptionally active chromatin—euchromatin—allows access of appropriate transcription factors to localize sequence-specific regions on target genes (Fig. 1). Once bound to their cognate DNA sequences, the transcription factors recruit coactivator proteins with histone acetylase activity to further loosen the chromatin structure and modify the nucleosome structure, allowing sliding of the chromosomal DNA. The additional recruitment of several basal transcription factors then gives rise to the formation of the preinitiation complex at the transcription start site and assists the RNA polymerase II during the synthesis of the hnRNA (Fig. 1). Control of chromatin organization and binding to *cis*-regulatory sequences represent the major check points in transcriptional regulation.

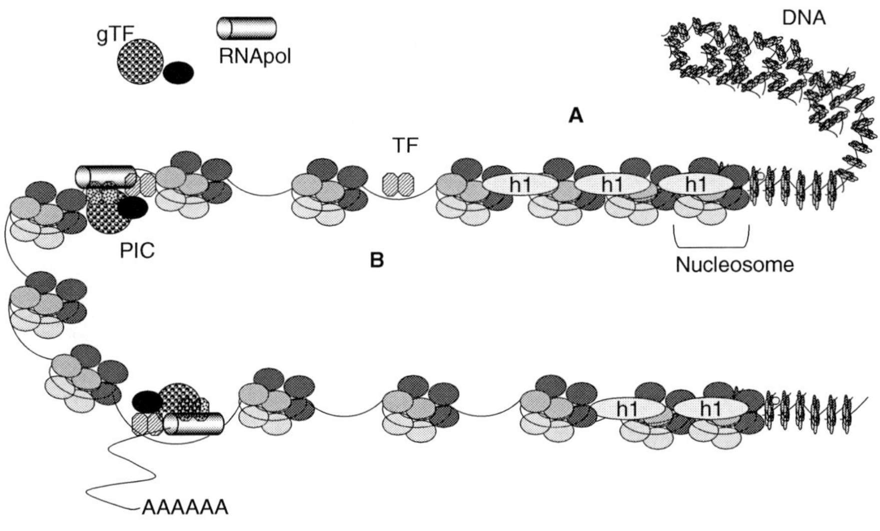

Fig. 1 Schematic representation of transcriptional activation of gene expression. **A.** Chromatin remodeling enzymes and histone acetylases cause decondensation of the DNA structure at the level of nucleosomes. **B.** Recruitment of transcription factors (TF) to specific accessible DNA-binding sites attracts coactivator and general transcription factors (gTF) to assist in the formation of the preinitiation complex (PIC) and transcription by RNA polymerase II (RNApol)

Hormetic Signaling Pathways

The application of genome-wide approaches has begun to provide a global view of gene expression responses to many different stress conditions (Murray et al., 2004; Kultz, 2005; Bahn et al., 2007). Signaling pathways that have began to emerge as common denominators between various insults include the Nrf2/ARE, FOXO, HSF, and NF-κB pathways.

Nuclear Factor–Erythroid 2p45 (NF-E2)–Related Factor (Nrf2)/Antioxidant Response Element (ARE) Signaling Pathway

Over several hundred million years of evolution, cells and organisms have developed a system of antioxidants and phase II detoxifying enzymes to protect them against the nefarious effects of electrophiles and omnipresent reactive oxygen species. Several studies have led to the identification of a common *cis*-acting enhancer element, 5′-GTGACnnnGC-3′, known as the antioxidant responsive element (ARE) or electrophile responsive element (EpRE), within the 5′ flanking regulatory regions of antioxidant and detoxifying genes (Surh, 2003; Shen et al., 2005). Among the ARE-responsive genes are glutathione-*S*-transferase (GST), glutathione reductase, epoxide hydrolase, aldehyde reductase, UDP-glucuronyl transferase (UGT), heme oxygenase-1 (HO1), NADP(H):quinone oxidoreductase (NQO1), thioredoxin, and γ-glutamylcysteine synthetase (γ-GCS) (Surh, 2003). The transcription factor that binds to the ARE consensus sequence and regulates the activation of ARE-target genes is Nrf2.

Nrf2, Keap1, and Regulation of the ARE Pathway

Nrf2 was isolated in 1994 as a factor binding to the NF-E2 sequence of the β-globin locus promoter region (Moi et al. 1994). It belongs to the "cap 'n' collar" (CNC) subfamily of the basic leucine zipper transcription factors (Zhang, 2006; Kensler et al., 2007). The family includes the closely related proteins p45 NF-F2, Nrf1, Nrf2, and Nrf3, plus two distantly related transcriptional repressors, Bach1 and Bach2 (Zhang, 2006; Kensler et al., 2007). Like other leucine zipper proteins, Nrf2 cannot bind to the ARE consensus sequence as a monomer or homodimer and must heterodimerize with members of the small Maf protein family for efficient DNA binding and transactivation activity (Itoh et al., 1997). Beside sMaf proteins, in vitro studies have shown that Nrf2 may form heterodimers with c-Jun and ATF4 (Venugopal and Jaiswal, 1998; He et al., 2001). Although the interaction with these proteins may be important in certain conditions (Venugopal and Jaiswal, 1998; He et al., 2001), their significance in vivo is unclear. Very little is known about the mechanisms regulating Nrf2 dimerization, the binding to the DNA, and the transactivation activity. The interaction with Bach 1 and Bach 2 has been shown to downregulate the expression of HO-1 and ARE-dependent reporter gene expression (Sun et al., 2002; Muto et al., 2002), leading to the assumption that they are Nrf2 transcriptional repressors. On the other hand, Nrf2 is able to bind the coactivator CREB-binding protein (CBP or p300), which in turn strongly enhances its transactivation activity (Katob et al., 2001). The best-understood Nrf2 regulatory mechanism is the interaction with the negative regulator Kelch-like ECH-associated protein 1 (Keap1) (Zhang, 2006; Kensler et al., 2007). Keap1 contains two major domains, an N-terminal broad complex/tramtrack/bric-a-brac (BTB) domain, and a C-terminal double-glycine repeat domain (DGR), separated by an intervening region. The DGR domain is essential for binding to the actin cytoskeleton and to the N-terminal of Nrf2 (Zhang, 2006; Kensler et al., 2007). The BTB domain is required for Keap1 homodimerization and for interaction with the Cullin3/Rbx1 E3 ubiquitin ligase (Kobayashi et al., 2004). Based on structural data, it is believed that in the absence of activators, Keap1 homodimers bind Nrf2 and actin, causing the retention of Nrf2 in the cytoplasm (Zhang, 2006; Kensler et al., 2007). Through the binding to Cul3/Rbx1 E3 ligase, Nrf2 is targeted for ubiquitination and subsequent proteasomal degradation (Kobayashi et al., 2004) (Fig. 2). On exposure to electrophilic compounds or reactive oxygen species, Nrf2 dissociates from Keap1, thus eluding degradation, and translocates in the nucleus, where it mediates the activation of ARE-target genes (Zhang, 2006; Kensler et al., 2007) (Fig. 2).

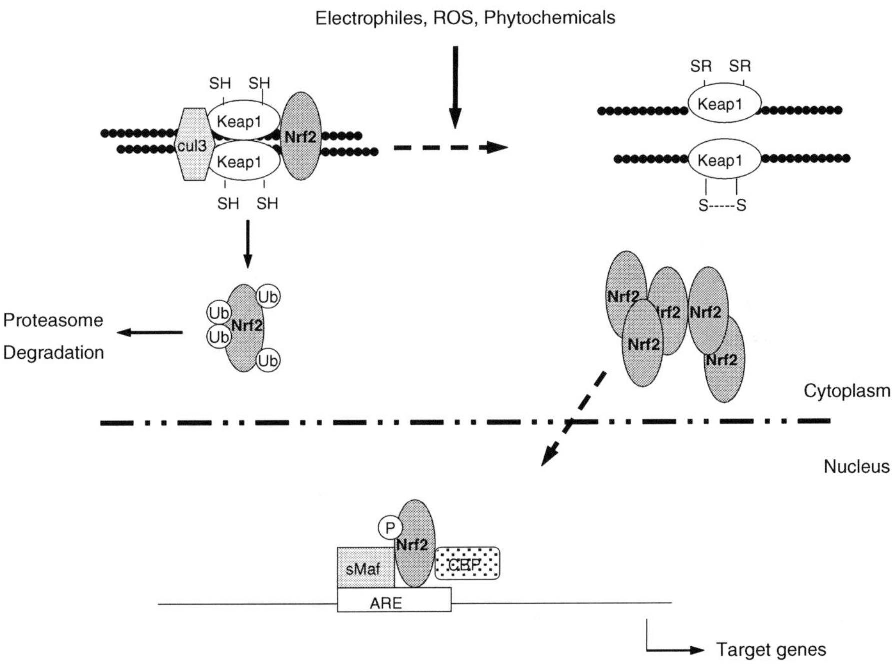

Fig. 2 Mechanism of induction of Nrf2-dependent genes. Under basal conditions, Nrf2 is bound to Keap1, retained in the cytosol, and targeted for proteasomal degradation via association with Cullin3/Rbx1–E3 ubiquitin ligase (Cul3). Inducers cause the release of Nrf2 from Keap1, allowing escape from degradation and nuclear relocalization. In the nucleus, Nrf2 binds to its cognate ARE site in association with small Maf protein members (sMaf), together with the coactivator protein (CBP), and regulates the expression of target genes

Because Keap1 is highly enriched in cysteine residues, it is believed to be an oxidative/electrophilic sensor. Many of the 27 cysteine residues are reactive and potential target sites for a direct interaction with electrophilic compounds (Nguyen et al., 2004; Kobayashi and Yamamoto, 2005). Cell-free experiments demonstrated that four cysteine residues located in the intervening region—Cys257, Cys273, Cys288, and Cys297—are extremely reactive and a direct target of certain phase II inducers, leading to Keap1/Nrf2 dissociation (Dinkova-Kostova et al., 2002). In addition, Cys273 and Cys288 are required for ubiquitylation and subsequent degradation by the proteasome (Zhang and Hannink, 2003). The importance of Keap1 in the regulation of Nrf2 stability is substantiated by observations in vivo. Keap1 knockout mice survive only 3 weeks after birth (Wakabayashi et al., 2003). Of interest, liver and embryonic fibroblasts from these mice exhibit increased expression of phase II detoxifying enzymes and constitutively higher levels of nuclear Nrf2 (Wakabayashi et al., 2003).

Hormetic Inducers of the Nrf2/ARE Pathway

Together with genes for antioxidants and phase II detoxification enzymes, Nrf2 has been shown to regulate genes involved in cell growth and apoptosis, inflammation, and the ubiquitin-mediated degradation pathway (Kwak et al., 2003; Lee et al., 2003a; Cho et al., 2005). Several in vitro and in vivo studies had demonstrated that Nrf2 is fundamental for the defense against reactive oxygen species and the pathogenesis of lung, hepatic, and neurodegenerative diseases. A better understanding of the functions, targets, and inducers of Nrf2/ARE-mediated gene expression has been achieved with the generation of Nrf2-deficient mice (Ramos-Gomes et al., 2001). Loss of Nrf2 decreases constitutive and inducible target gene expression, enhancing the sensitivity of Nrf2-deficient mice to carcinogenesis and oxidative stress (Chan et al., 2001; Ramos-Gomez et al., 2001; Cho et al., 2002). Nrf2-deficient mice have a significantly higher risk of benzo[a]pyrene-induced gastric neoplasia than do wild-type mice, due to the reduction of constitutive hepatic and gastric activities of GST and NQO1 (Ramos-Gomes et al., 2001). Hyperoxia-induced expression of NQO1, GST, UGT, glutathione peroxidase-2 (GPx2), and HO1 is significantly lower in Nrf2-deficient mice (Cho et al., 2002). Lack of Nrf2 sensitizes neurons and astrocytes to oxidative stress by decreasing constitutive and inducible protective genes (Lee et al., 2003a; Lee et al., 2003b).

In *Caenorhabditis elegans* the homolog of Nrf2, SKN1, integrates stress tolerance responses with energy metabolism homeostasis, regulating lifespan (Tullet et al., 2008). Recently Tullet et al. (2008) showed that the insulin/IGF1 pathway directly regulates SKN1, with unique functions in different tissues. Restricted expression of SKN1 in ASI neurons (putative neuroendocrine cells) mediates calorie restriction–induced extension of lifespan, suggesting that its role in these neurons is to tune the systemic responses to nutrition (Bishop and Guarente, 2007b). It is intriguing that intestinal SKN1 responds to environmental stress by promoting the expression of phase II genes, and its prolongevity effect is distinct from its ASI-mediated role in calorie restriction (An and Blackwell, 2003; Bishop and Guarente,

2007). It is possible that the stress and tissue specific sensitivity and differential function of SKN1/Nrf2 observed in worms are conserved in higher vertebrates. Notably, we recently observed that in ARE-hPAP reporter mice, starvation causes a significant upregulation of hPAP Nrf2-driven expression in liver but not in the cerebral cortex (Fig. 3). In contrast to the situation in worms, in mice Nrf2 seems to mediate calorie restriction anticarcinogenic properties, but apparently it is dispensable for its prolongevity benefits (Pearson et al., 2008). The discrepancy observed between worms and mice on how calorie restriction extends lifespan could be due to different mechanisms of IGF1/insuling-signaling regulation of Nrf2.

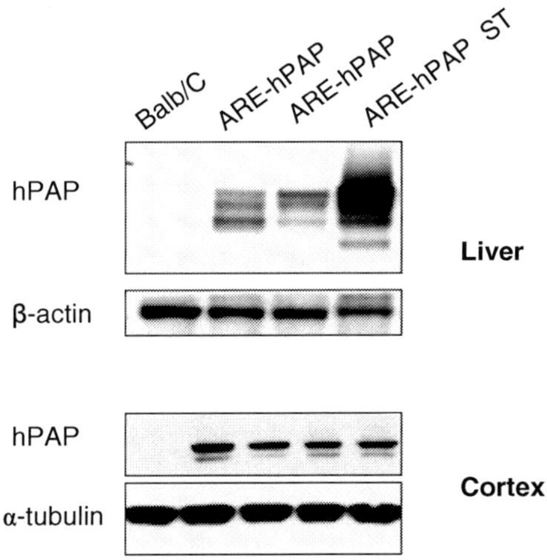

Fig. 3 Tissue-specific effects of starvation on Nrf2 activation. ARE-driven human placental alkaline-phosphatase mice (ARE-hPAP) were subjected to 2 days of starvation (ST). Levels of hPAP were measured by Western blot analysis in extracts from liver and cerebral cortex using actin and tubulin as loading controls

Several phytochemicals with beneficial health effects have been shown to mediate Nrf2 activity. Sulforaphane is an isothiocyanate present in high amounts in broccoli sprouts and cruciferous vegetables (Myzak and Dashwood, 2006). Sulforaphane's anticarcinogenic and protective effects are, at least partially, due to its ability to activate the Nrf2/ARE pathway. Indeed, in vivo gene-expression analysis comparing wild-type mice and Nrf2-null mice led to the identification of specific sulforaphane upregulated ARE-dependent target genes (Thimmulappa et al., 2002). Sulforaphane protects cultured neurons against oxidative stress (Kraft et al., 2004) and dopaminergic neurons against mitochondrial toxins (Han et al., 2007). Administration of sulforaphane to mice can protect photoreceptors against degeneration in a retinal degeneration model (Kong et al., 2007). From a mechanistic point of view, sulforaphane acts at different levels of the Nrf2 pathway. It suppresses Nrf2 proteasomal degradation, leading to a prolonged half-life and transcriptional activity (Jeong et al., 2005), and directly covalently binds the thiol groups in the inhibitor Keap1, thus causing the release of Nrf2 and its subsequent nuclear relocalization (Dinkova-Kostova et al., 2002). The antioxidant carnosol induces HO1 expression and activates ARE- reporter activity in PC12 cells in a phosphatidylinositol 3 kinase (PI3K)– and AKT-dependent fashion (Martin et al., 2004). Treatment of endothelial cells with epigallocatechin-3-gallate (EGCG), the major constituent of green tea, increases Nrf2 nuclear levels and upregulates HO1 expression (Wu et al., 2006). Resveratrol increases the activities of catalase, superoxide dismutase, glutathione peroxidase, NQO1, and GST and upregulates Nrf2 and induces its translocation to the nucleus (Hsieh et al., 2006; Rubiolo et al., 2008). In addition, to the phytochemicals cited previously, several other natural compounds have been identified as inducers of the Nrf2/ARE pathway, including quercetin (Tanigawa et al., 2007), curcumin (Kang et al., 2007), phenethyl isothiocyanate (Son et al., 2008), and chalcone (Son et al., 2008).

Forkhead Box O (FOXO) Transcription Factors

The mammalian orthologs of *C. elegans* DAF-16 [a forkhead/winged-helix transcription factor, box O (FOXO)] transcription factors FOXO1, FOXO3a, FOXO4, and FOXO6 function as tumor suppressors and energy metabolism rheostats and are essential proteins for animals to reach their genetic maximum lifespan (Libina et al., 2003). The expression of FOXO-regulated genes can be controlled by any of the transcription factors, and specificity is achieved either by expression-pattern or by isoform-specific regulation. We will therefore use the general term FOXO, unless otherwise specified. The FOXO protein family is characterized by an N-terminal 100-base DNA-binding domain, or FKH domain, containing three major α-helices and two large, wing-like loops (Clark et al., 1993), a C-terminal trans-activating domain, and nuclear localization and exclusion sequences. Although the molecular basis of the DNA-binding specificity is poorly understood, it appears that both the α-helices and winged loops of the FKH region are important for the interaction of the FOXO monomers with the DNA (Clark et al., 1993; Boura et al., 2007). High-affinity DNA-binding studies have led to the identification of the consensus sequence FOXO-recognized element (FRE) as (G/C)(T/A)AA(C/T)AA (Furuyama et al., 2000; Gilley et al., 2003).

Through systematic bioinformatic approaches and direct experimentation, putative and functional FRE sites have been identified in the promoter of hundreds of genes. In fact, FOXO factors are downstream targets of many anabolic growth factors (e.g., insulin, insulin-like growth factor 1 [IGF1]) and cellular nutrition and oxidative stress response pathways that coordinate a wide range of cellular functions, depending on the type and strength of the stimulus. FOXO-regulated responses include increased metabolic efficiency via gluconeogenesis (e.g., glucose-6-phosphatase, PEPCK, SCPx, AgRP, NPY), enhanced stress resistance (e.g., catalase, MnSOD), DNA repair (e.g., GADD45, DDB1, sestrin), neuropeptide secretion (e.g., *C. elegans* SCL-1), atrophy/autophagy (e.g., mTOR), cell cycle arrest (e.g., p27, GADD45, p130Rb2, cyclin G2), and apoptosis (e.g., Bim-1, FasL, BCL-6, PUMA) (Vogt et al., 2005; Hansen et al., 2008). FOXO functions are controlled through a complex interaction of three posttranslational modifications: phosphorylation, acetylation, and ubiquitination. Some modifications control subcellular localization, whereas others modulate FOXO transactivation properties.

The nuclear/cytosol shuttling of FOXO appears to be a key regulator of cellular anabolic (i.e., FOXO export) and catabolic/apoptosis (i.e., FOXO import) transcriptional signaling and is tightly controlled through phosphorylation. Two main classes of stimuli that trigger FOXO phosphorylation with opposite effects on subcellular localization are developmental and environmental stress signals. Phosphorylation of FOXO in response to growth factors such as IGF1, epidermal growth factor, or nerve growth factor results in the export of FOXO from the nucleus and abrogation of its transcriptional activity (Fig. 4). The kinases involved include AKT and serum and glucocorticoid inducible kinases (SGKs), both activated trough the PI3K pathway, dual-specificity tyrosine phosphorylation–regulated kinase 1 (DYRK1), and casein kinase 1 (CK1) (Biggs et al., 1999; Brunet et al., 1999; Rena et al., 2002;Vogt et al., 2005; Hansen et al., 2008). The phosphorylation of nuclear FOXO reduces its ability to bind DNA and recruit coactivators such as p300/CBP, thereby decreasing its transactivation potential (Tsai et al., 2003). Nuclear phosphorylated FOXO becomes a target for 14-3-3 proteins and CRM1/Ran-dependent export from the

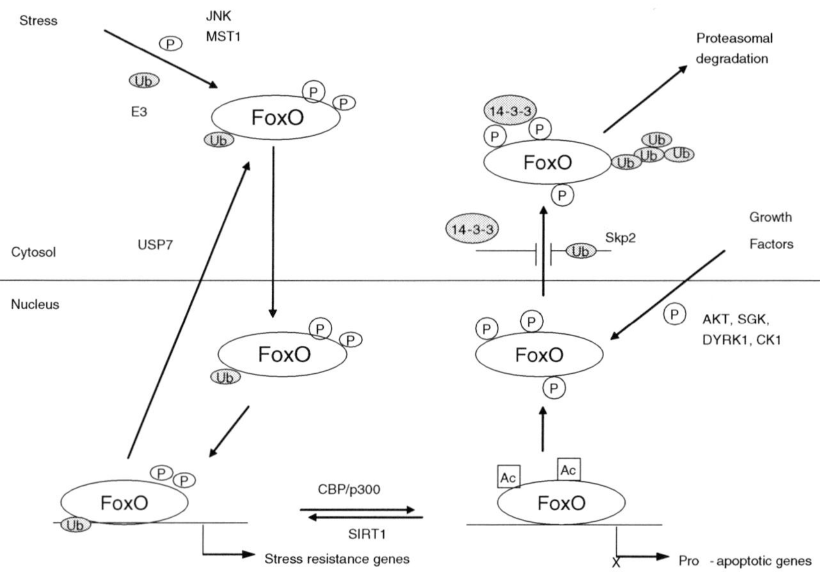

Fig. 4 FOXO posttranslational regulatory mechanisms. Phosphorylation of FOXO in response to growth factor stimulation causes its relocalization in the cytosol through 14-3-3 and CRM1-dependent export mechanisms. The cytosolic retained FOXO is then targeted for polyubiquitination and 26S proteasomal degradation. On the other hand, stress-induced monoubiquitination and/or phosphorylation promote FOXO dissociation from 14-3-3 proteins and translocation to the nucleus. FOXO transcriptional activity can be modulated in the nucleus through reversible acetylation. Deacetylation by SIRT1 seems to enhance transcriptional activity of stress resistance genes while inhibiting proapoptotic genes

nucleus (Brunet et al., 2002). In addition to nuclear expulsion, the interaction with 14-3-3 proteins prevents reentry of FOXO into the nucleus by masking the nuclear localization signal (Rena et al., 2001). The cytosolic retained phosphorylated FOXO is then polyubiquitinated (i.e., ligase Skp2) and targeted for 26S proteasomal degradation (Plas and Thompson, 2003; van der Heide and Smidt, 2005) (Fig. 4). The polyubiquitination and turnover of FOXO's have been shown to be an irreversible, slow, and constitutive process (Plas and Thompson, 2003).

On the other hand, phosphorylation of FOXO by JNK (Essers et al., 2005) or mammalian Ste20-like kinase (MST1) (Lehtinen et al., 2006) in response to genotoxic or oxidative stress promotes the dissociation from 14-3-3 proteins and translocation into the nucleus, whereas FOXO upregulates antioxidants such as MnSOD and catalase (Kops et al., 2002), as well as heat-shock proteins (Wang et al., 2005) (Fig. 4). In contrast to growth factor–induced degradation, oxidative stress induces a reversible monoubiquitination of cytosolic FOXO, which does not affect its stability but instead induces its nuclear relocalization and transcriptional activity (van der Host et al., 2006). Although it is unclear which E3 ligase catalyzes the import process, the deubiquitination of FOXO and subsequent nuclear export are mediated by the herpes virus–associated ubiquitin-specific protease (USP7) (van der Host et al., 2006). Notably, the effects of the stress-initiated pathway appears to prevail over the effects of growth factors (Brunet et al., 2004; Essers et al., 2004).

In the nucleus, FOXO proteins regulate target genes through multiple modes of action. FOXO can recruit coactivators or cooperating transcription factors, which can either enhance (e.g., β-catenin) (Essers et al., 2005) or repress (e.g., PPARγ) (Dowell et al., 2003) their transcriptional activity. Alternatively, FOXO may repress transcription by competing with other factors for a common binding site in a gene promoter (Kitamura et al., 2002). Finally, FOXO can act as coactivator or corepressor for other transcription factors, thereby regulating promoters lacking FRE-binding sites (Glauser and Schlegel, 2007). Although phosphorylation of FOXO's has been shown to decrease their transactivation potential by inhibiting their interaction with coactivators (Tsai et al., 2003), nuclear interactions and transcriptional properties of FOXO's are mainly modulated by reversible acetylation. The acetyl transferase p300/CBP binds and acetylates FOXO proteins, although the net consequences (activation vs. repression) are controversial (van der Heide and Smidt, 2005). Deacetylation of FOXO is mediated by the mammalian SIR2 ortholog SIRT1 (Daitoku et al., 2004; Motta et al., 2004). The interaction between SIRT1 and FOXO factors occurs in response to oxidative stress (Brunet et al., 2004; Motta et al., 2004). In *C. elegans* genetic manipulations show that sir2.1 increases longevity through deacetylation and activation of DAF16/FOXO (Tissembaum and Guarente, 2001). In mammals the effects of SIRT1 on FOXO functions are much more complex and vary depending on the target genes. In general, it appears that SIRT1 promotes the transcription of FOXO target genes involved in stress resistance while decreasing the transcription of proapoptotic genes (Greer and Brunet, 2005). It is possible that, in contrast to *C. elegans*, increased longevity in mammals may be achieved by using SIRT1 to block FOXO's induction of the apoptosis pathway while maintaining the response to increase repair and protection mechanisms during stress conditions like calorie restriction (van der Host and Burgering, 2007).

FOXO, Oxidative Stress, and Longevity

In recent decades FOXO's have emerged as master signaling integrators translating environmental stimuli into dynamic gene expression programs that influence proliferation, apoptosis, metabolism, stress resistance, and longevity (van der Host and Burgering, 2007). FOXO transcription factors are activated by environmental stressors such as starvation, overcrowding (Gross et al., 2008), oxidative stress (Essers et al., 2005), and gamma irradiation (Tran et al., 2002). In addition, several studies have shown that polyphenolic phytochemicals from tea, such as epigallocatechin-3-gallate (EGCG), black tea theaflavins, theaflavin 3-*O*-gallate, theaflavin 3′-*O*-gallate, theaflavin 3,3′di-*O*-gallate, and thearubigins, activate FOXO1a and have many health benefits, such as inhibiting angiogenesis and aiding in the management of diabetic retinopathy, rheumatoid arthritis, psoriasis, cardiovascular diseases, and cancer (Cameron et al., 2008; Shankar et al., 2008).

Are FOXO's longevity determinant proteins? In invertebrates, calorie restriction–induced increase in lifespan and health is linked to the insulin/IGF1 signaling pathway. The underlying mechanism(s) require (1) a slower accumulation of products from the activation of the insulin pathway; (2) a higher activity of stress response survival genes; or (3) both. In *C. elegans* low insulin/IGF1 signaling (e.g., insulin/IGF1 receptor homolog DAF-2 mutants, or calorie restriction) signals through a conserved PI3 kinase/AKT pathway, inhibiting DAF-16 (i.e., FOXO). Daf-16 knockout mutants live an average or shorter adult lifespan even under starvation conditions (Libina et al., 2003). DAF-2 leaky mutants have dauer-like phenotypes and remain youthful and active twice as long as normally developed adult worms (Libina et al., 2003). Under normal growth conditions, insulin/IGF1/AKT signaling phosphorylates and inhibits DAF-16, resulting in a normal lifespan. However, environmentally activated (e.g., starvation, overcrowding) nuclear DAF-16 upregulates a wide variety of genes, including cellular stress response, antimicrobial, and metabolic genes and has been postulated to downregulate specific life-shortening genes (Libina et al., 2003).

The insulin/IGF1/AKT/FOXO pathway is virtually identical in worms, flies, and mammals, and therefore has been conserved throughout evolution (Burgering and Kops, 2002). This does not imply that FOXO functions are also identically conserved. It is likely that the contributions of FOXO in the regulation of lifespan are organism dependent. In a short- lived organism like *C. elegans*, which, except for the reproductive cells, consist only of postmitotic cells, FOXO functions are mainly channeled on regulating metabolic stress to survive food shortage. In organisms with a longer lifespan and a soma with multiple proliferative cell types, the necessity of genotoxic stress resistance for survival and prevention of cancer is greater. Thus metabolic stress resistance, and FOXO, will have a different impact on lifespan. Indeed, although the role of DAF-16/FOXO in calorie restriction lifespan extension through activation of the dauer pathway is striking in nematodes, the effect in other species is less straightforward. In *Drosophila melanogaster*, ablation of dFOXO does not perturb the increase in longevity induced by calorie restriction, and overexpression of dFOXO minimally (i.e., 10% or less) increases average and maximum lifespan (Giannakou et al., 2008; Min et al., 2008). However, IGF1 receptor knockout mice show a 30% increase in maximum lifespan, up to 40% after calorie restriction, suggesting a possible involvement of FOXO (Bluher et al., 2003; Holzenberg et al., 2003).

Based on the hormetic concept, it is possible that the periodic induction of FOXO-dependent genes by mild pulses of hormetic stressors implements the cleanup of damage to below basal levels and increases antioxidant protection, resulting in enhanced cellular function and performance. On the other hand, chronic activation is likely to negatively impact cell homeostasis, leading to disease. Notably, deletion of one allele of FOXO1 (*FOXO1* $^{+/-}$, heterozygous knockout) in the insulin receptor–deficient diabetic mouse model (i.e., *Insr* $^{+/-}$) reverses the diabetic phenotype as the result of a 75% decrease in FOXO1-induced glucose-6-phosphatase activity, whereas overexpression of FOXO1 results in diabetes (Nakae et al., 2002). Similarly, deletion of one allele of FOXO1 rescues beta cell proliferation in *βpdky* $^{-/-}$ mice, alleviating their diabetic phenotype (Hashimoto et al., 2006).

The Nuclear Factor-κB Pathway

Nuclear factor κB (NF-κB) is one of the most-studied and best-understood inducible transcription factors. The NF-κB/Rel family consist of five proteins—RelA (p65), cRel, RelB, p50, and p52—sharing a 300–amino acid N-terminal Rel homology domain (RHD) responsible for DNA binding, nuclear translocation, interaction with IκBs, and homo- and heterodimerization. The transactivation domain (TD) necessary for the positive regulation of gene expression is present only in p65, cRel, and RelB. Because of the lack of a transactivation domain, p50 and p52 homodimers usually act as transcriptional repressors of κB-responsive promoters, unless they interact with other TD-containing NF-κB members. NF-κB dimers are typically bound to an ankyrin repeat enriched IκB protein family member, for example, IκBα, IκBβ, IκBϵ, IκBγ, IκBζ, Bcl3, p100, and p105. Although historically considered as inhibitors of NF-κB activity, the recent advances in understanding the functions of IκBs indicate that they act as cofactors of NF-κB, able to either inhibit or promote (i.e., Bcl3 and IκBζ) κB-driven transcription.

The NF-κB dimers bind to promoter and enhancer regions containing the κB consensus sequence 5′-GGGRNWYYCC-3′ (N = any base; R = purine; W = adenine or thymine; Y = pyrimidine) (Baeuerle, 1991; Hoffmann et al., 2006). The degenerate nature of the κB site sequence, together with the different binding preferences of the various dimers, the use of various IκBs, and the posttranslational modification regulating the interactions with coactivators and corepressors, accounts for the large list of NF-κB target genes (Gilmore 2008).

The prototypical and best-studied NF-κB signaling pathway is the so-called "canonical" or "classical" pathway, involving IκBα/p65/p50 complexes. In resting conditions, despite steady-state localization that appears almost exclusively cytosolic, the IκBα/p65/p50 complexes continuously shuttle between the nucleus and the cytosol (Ghosh and Karin, 2002) due to the presence of an exposed nuclear localization signal (NLS) of p50 and a nuclear exclusion signal (NES) in IκBα. Following stimulation by one of the many inducers (Gilmore, 2008), the activated complex IκB kinase kinase (IKK) phosphorylates IκBα on Ser32 and Ser36, targeting the protein for K48-linked polyubiquitination by the βTrCP-containing Skp1-Culin-Roc1/rbx1/Hrt-1-F-box ($SCF^{\beta TrCP}$) E3 ubiquitin ligase complexes and E2 UbcH5. Once ubiquitinated, IκBα is rapidly degraded through the proteasome, thus unmasking the NLS of p65 and pushing the balance between NLS and NES in favor of the nuclear localization. Among others genes, nuclear NF-κB drives the expression of IκBα, generating a negative feedback loop essential for the proper termination of the signaling response. The "alternative" or "noncanonical" pathway appears to be relevant only in the immune system responding to a very limited subset of stimuli, such as lymphotoxin B, CD40 ligand, and B-cell activating factor, and is specific for p100/RelB dimers (Senftlenben et al., 2001). Engagement of the receptor activates NF-κB–inducing kinase (NIK), which in turn directly phosphorylates IKKα homodimers. The subsequent phosphorylation of p100 at Ser866, Ser870, and Ser872 leads to recruitment of $SCF^{\beta TrCP}$ E3, polyubiquitination of Lys855, and partial degradation or processing to generate p52/RelB dimers (Hayden and Ghosh, 2008).

NF-κB as a Hormetic Transducer of Exercise

Regular exercise of moderate to high intensity (training) is known to be beneficial in preventing diseases and improving physiological functions. The benefits are not limited to the skeletal muscle and heart (Radak et al., 2005; Haskell et al., 2007), but extend to other organs (Radak et al., 2001; Cotman et al., 2007). For example, in rats, regular swimming training significantly decreases oxidative damage to lipids, proteins, and DNA not only in skeletal muscle but also in brain (Radak et al., 1999), significantly improving cognitive functions (Radak et al., 2001).

During exercise, contracting muscles consume as much as 100-fold more oxygen than resting muscles, generating massive amounts of reactive oxygen species (ROS) through mitochondrial electron transport chain, adenine nucleotide catabolism, lipooxigenase, and NADPH oxidase (Davies et al., 1982; Alessio et al., 1988). With training, the repeated exposure to ROS leads to antioxidant adaptation, increasing the resistance of the organism to oxidative stress. Exercise has been shown to upregulate the activity of MnSOD (Hollander et al., 2000), inducible nitric oxide synthase (Gomez-Cabrera et al., 2005), and γ-glutamylcysteine synthase (Leeuwenburgh et al., 1997) in a transcription-dependent fashion. A shared factor among these enzymes is their responsiveness to the NF-κB signaling pathway. Exercise has been shown to upregulate NF-κB activity in rodents, acting at different points of the signaling cascade (Kramer and Goodyear, 2007). Studies in rodent models have demonstrated that the NF-κB content and activation patterns are muscle fiber type dependent (Hollander et al., 2000; Atherton et al., 2004; Durham et al., 2004). For example, basal levels of NF-κB in soleus muscle are two- to threefold higher than in extensor digitorum longus (Atherton et al., 2004). These differences are likely attributable to differences in the oxidative and/or metabolic status of the muscle fiber types. NF-κB binding is significantly elevated in rat skeletal muscle after an acute bout of exercise, leading to upregulation of MnSOD mRNA levels and protein content in the muscle (Hollander et al., 2001).

One hour of treadmill exercise in rats induces activation of NF-κB in the soleus (type 1) and gastrocnemius (type 2a) muscle with increased IKKα/β phosphorylation (Ho et al., 2005). In mice, NF-κB activity is decreased in diaphragm fiber bundles after a tetanic exhaustive stimulation; however, 12 days of training increases NF-κB binding activity in soleus muscle (Durham et al., 2004). Studies in humans revealed that peripheral blood lymphocytes from marathon runners have ROS-dependent increased levels of activated NF-κB compared to their preexercise status (Gomez-Cabrera et al., 2006). On the other hand, in vastus lateralis muscle biopsies a fatiguing resistance contraction transiently reduces the NF-κB preexercise binding levels (Durham et al., 2004). Therefore it appears that the nature of the exercise performed, in terms of intensity, duration, and frequency, as well as the time of recovery between exercises, influences the activation status of NF-κB and the balance between muscle adaptation and damage. Intermittent activation of NF-κB followed by sufficient recovery, as during training, leads to efficient muscle repair and antioxidant upregulation, allowing the body to better cope with subsequent mechanical and oxidative stress (Fig. 5). On the other hand, persistent activation of NF-κB with limited recovery time impairs tissue regeneration and exacerbates muscle inflammation, favoring muscle wasting (Fig. 5). This scenario is supported by the observation that mice carrying a muscle specific activation of IKKβ, thus showing high and persistent levels of muscular NF-κB activity, had profound sarcopenia, increased proteolysis, and overall a cachectic phenotype (Cai et al., 2004). Preventing the activation of NF-κB by crossing the mice with muscle-specific IκBα superrepressor mice reverted the phenotype, thus strongly suggesting that NF-κB chronic activation directly mediated cachexia (Cai et al., 2004). In addition to the canonical pathway, both Bcl3 and p105/p50 are also implicated in muscle atrophy (Hunter and Kandarian, 2004). Nfkb1 (p105/p50) and Bcl3 knockout mice showed resistance to unloading-induced atrophy in soleus and plantaris muscles (Hunter and Kandarian, 2004). Again a fiber-specific pattern was observed; fast fibers were more resistant in both types of mice, and only $Bcl3^{-/-}$ slow fibers had decreased atrophy.

Heat-Shock Factor Pathway

The heat-shock response was first discovered in 1962 by F. Ritossa, who detected a new puffing pattern on heat-shock treatment in the polytene chromosomes of the fruitfly *Drosophila buschii* (Ritossa, 1962). Subsequent studies elucidated the nature of the induced RNAs and proteins at the molecular level and led to the isolation and characterization of a highly conserved set of proteins, the heat-shock proteins (HSPs) (Lindquist, 1986; Lindquist and Craig, 1988). It is now clear that the protection and recovery from cellular damage afforded by the HSPs is elicited by a

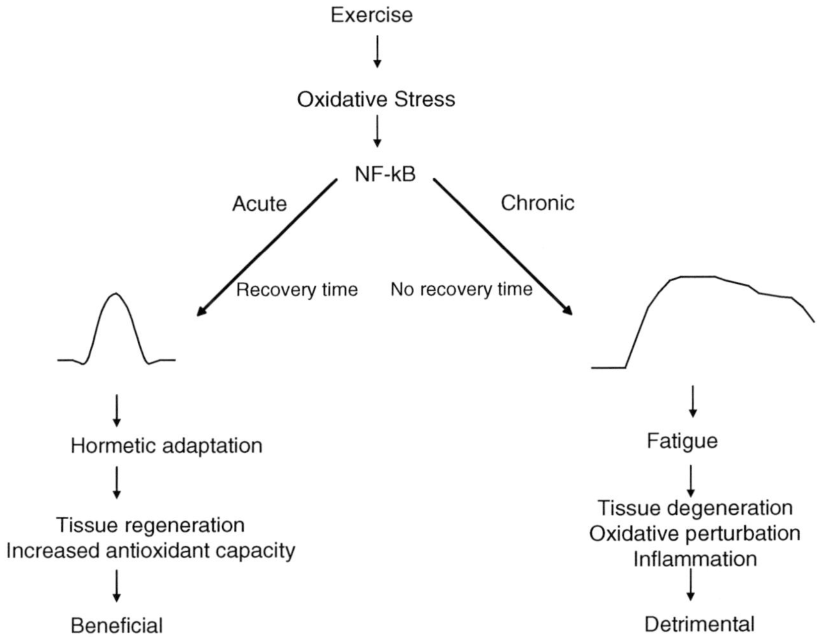

Fig. 5 NF-κB is transiently activated in muscle cells during exercise. Allowing sufficient recovery time between bouts of exercise (e.g., training) promotes the expression of beneficial proteins conferring to the organism increased resistance to subsequent stressors (hormetic adaptation). In contrast, chronic activation of NF-κB or insufficient recovery time results in persistent muscle inflammation with tissue degeneration

large number of acute and chronic conditions besides elevated temperatures, such as heavy metals, chemical toxicants, oxidative stress, infection, hypoxia acidosis, energy depletion, and exercise (Kregel, 2002; Westerheid and Morimoto, 2005).

The HSP family is organized by molecular size, and includes many proteins functioning as proteases and molecular chaperones to guide conformational states critical in the synthesis, folding, translocation, assembly, and degradation of proteins (Westerheid and Morimoto, 2005). The heat-shock response is regulated at the transcriptional level by the activities of the heat-shock transcription factors (HSFs) (Pirkalla et al., 2001). Humans express three HSFs—HSF1, HSF2, and HSF4—each characterized by two isoforms—a and b (Wu, 1995; Pirkalla et al., 2001). HSF1 is the best characterized, and is the undisputed master regulator of the heat-shock response. HSF1-deficient mice, other than being about 20% smaller than wild-type mice, do not display particular organ abnormalities and live to late adulthood (Xiao et al., 1999). Although dispensable for growth and survival in controlled conditions, HSF1 absence completely abolishes the induction of HSPs in response to heat shock and other challenges, thus proving essential for survival following stress (Xiao et al., 1999).

The physiological functions of HSF2 and HSF4 are less clear. HSF2 appears to be mainly involved in development (Rallu et al., 1997), and, although historically considered stress insensitive, recently its isoform HSF2a was suggested to cooperate with HSF1 in stress-induced gene activation (He et al., 2003; Ostling et al., 2007). The two alternative splicing forms HSF4a and HSF4b differ by only 30 amino acids in the regulatory region yet have strikingly different functional properties when overexpressed in cells. HSF4a acts as an inhibitor of constitutive and inducible HSP expression (Nakai et al., 1997; Tanabe et al., 1999), whereas HSF4b is a transcriptional activator (Tanabe et al., 1999). The endogenous functions of both HSF4 isoforms are unknown. In general, HSF4 appears to be non responsive to stress yet to be needed for the maintenance of sensory organs such as lens (Fujimoto et al., 2004).

The HSF protein structure is well conserved (Pirkalla et al., 2001), with an N-terminal DNA-binding domain comprising a winged helix-turn-helix motif (Clos et al., 1990) (Fig. 6a) that interacts with the so-called heat-shock element (HSE) sequences present in the promoter of heat-shock protein genes. Two hydrophobic heptad regions (HR-A/B), characteristic for helical coiled-coil structures (i.e., leucine zippers), are located near the DNA-binding domain and are required for activation-induced HSF oligomerization. An additional hydrophobic repeat, HR-C, further downstream, is likely to mediate the suppression of HSF trimerization (Rabindran et al., 1993). The C-terminal transactivation domains appear to be regulated by a loosely defined regulatory domain (Green et al., 1995; Kline and Morimoto, 1997; Voellmy, 2004) (Fig. 6A).

In normal conditions HSF1 exists as an inert monomer in either the cytoplasmic or nuclear compartment. On exposure to a variety of stresses, HSF1 is derepressed, trimerizes, and accumulates in the nucleus. HSF1 trimers bind with high affinity to HSE cognate sequence consisting of multiple contiguous inverted repeats of the pentameric sequence NGAAN (N = any base) located in the promoter regions of target genes (Xiao et al., 1991). The minimal sequence required for HSF1 binding consists of two HSE elements, but three elements allow formation of the strongest HSF1–DNA complex. The genes encoding HSP70, HSP90, and sHSPs are also endogenously transcribed due to multiple basal factors or binding of low levels of active HSF1 (Westerheid and Morimoto, 2005).

Recent observations point to additional regulatory mechanisms. HSF1 can indeed be modified by phosphorylation and sumoylation. Sites of constitutive phosphorylation—Ser 303, Ser307, and Ser363—are important for the negative regulation of HSF1 and are targeted by GSK3, ERK1, and PKC, respectively (Knauf et al., 1996; Kline and Morimoto, 1997; Chu et al., 1998). In the cases of pSer 303 and pSer307, at least part of the inhibition is due to recruitment of 14-3-3 and nuclear export of HSF1 (Wang et al., 2003; Wang et al., 2004). Sites of inducible phosphorylation—Ser230, Ser326, and Ser419—promote HSF1 activity (Holmberg et al., 2001; Guettouche et al., 2005; Kim et al., 2005). Sumoylation on Lys298 after Ser303 phosphorylation has also been observed (Hietakangas et al., 2003), and appears to modulate HSF1 DNA binding (Anckar et al., 2006) and transcriptional activity (Hong et al., 2001). HSF1 is negatively regulated through interaction with

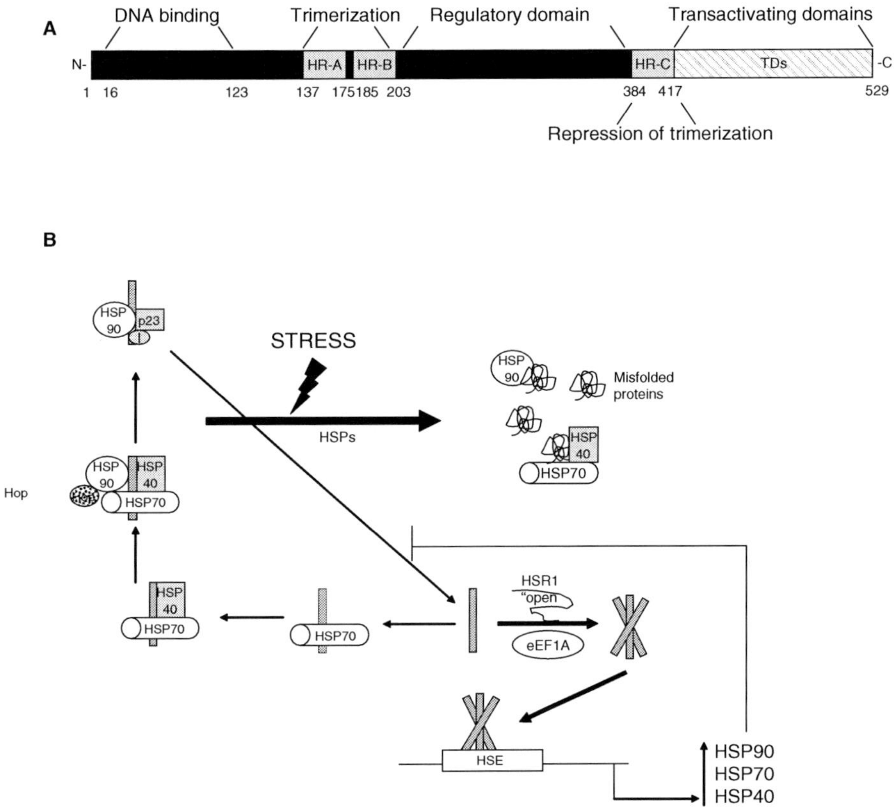

Fig. 6 HSF1 structure and stress-induced activation model. **A.** Structural domains of human HSF1α protein. Numbers refer to amino acid residues. HR, heptad repeat; TDs, transactivation domains. **B.** Schematic showing assembly of the inhibitory HSF complex. Under basal conditions, HSPs and cofactors bind sequentially to HSF1 monomers, preventing trimerization. This process is in a dynamic equilibrium with the handling of protein misfolding. Under stress conditions, the drastic augmentation of misfolded proteins shifts the equilibrium by sequestering HSPs and decreasing or interrupting the complex formation. The concomitant collapse of the cytoskeleton structure and blockage of mRNA translation render the elongation factor eEF1A available for interaction with the activated form of HSR1, thus allowing the formation of HSF1 trimers. HSF1 trimers bind to the cognate DNA sequence, driving the expression of heat-shock proteins, which can negatively feed back

HSP90 (Zou et al., 1998). The current negative feedback model implies that inactive, monomeric HSF1 undergoes a multistep assembly process (Fig. 6b). Initially HSF1 binds HSP70 and HSP40, and then HSP90 and the HSP90-binding protein Hop bind, causing the formation of a heterocomplex that includes HSP90, p23, and immunophilin (Hietakangas et al., 2003). These complexes are dynamic, such that when the cell is stressed, protein unfolding increases and the concentration of nonnative proteins increases. The nonnative proteins serve as substrate for HSP chaperones, other chaperons, and cofactors. Because of the competition between nonnative proteins and HSF1 oligomers, the rate of HSF1 heterocomplex assembly is reduced, releasing HSF1 for homotrimerization and binding to the DNA (Fig. 6B).

Recently an additional regulatory mechanism involving the noncoding RNA, HSR1, and translation elongation factor eEF1A was proposed (Shamovsky et al., 2006; Shamovsky and Nudler, 2008). In normal conditions HSR1 is present in an inactive "closed" conformation, whereas the majority of eEF1A is engaged in cytoskeleton maintenance (Negrutskii and El'skaya, 1998; Gross and Kinzy, 2005) and mRNA translation (Gross and Kinzy, 2005). During stress HSR1 changes to an "open" HSF1-activating conformation, whereas the collapse of cytoskeleton (Welch and Suhan, 1985) and general shutdown of protein synthesis (Panniers, 1994) release eEF1A. The depletion of chaperones prevents the formation of the inhibitory complex and eEF1A, and the formation of HSF1 trimers is either directly promoted by HSR1 and eEF1A or favored by the stabilizing interaction with them (Fig. 6B).

Nuclear HSF1 activity is further modulated through competitive equilibrium between HSE sites in the promoter of HSF1-responsive genes and satellite III repetitive sequences (nuclear stress bodies) on chromosomes 9, 12, and 15, where the majority of HSF1 accumulates (Jolly et al., 1997; Jolly et al., 1999; Denegri et al., 2002; Jolly et al., 2002). These satellite III sequences are composed of thousands of copies of redundant degenerate HSEs and are actively transcribed to generate noncoding RNA molecules whose functions remain to be established (Biamonti, 2004; Jolly et al., 2004).

Conclusions

Transcription factors play pivotal roles in hormetic responses of cells to many different types of intrinsic and environmental stressors. Typically, transcription factors are activated by stress-responsive kinases such as AKT, GSK3, MAP kinases, PKC, JNK, IKKs, and Ca^{2+}-sensitive kinases. Four transcription factors that mediate adaptive stress responses are Nrf2, FOXO's, NF-κB, and HSFs. These transcription factors target genes encoding several different types of cytoprotective proteins, including antioxidant enzymes (superoxide dismutases, glutathione peroxidase, etc.), antiapoptotic proteins (Bcl-2, inhibitor of apoptosis proteins, etc.), and protein chaperones (HSP70, HSP90, and others). Hormetic transcription factors respond to general types of metabolic and oxidative stress, but they may also respond to specific naturally occurring or man-made chemicals. For example, the phytochemicals curcumin and resveratrol activate Nrf2 and FOXO's, respectively. A better understanding of the mechanisms by which transcription factors respond to cellular stress and the elucidation of behavioral, dietary, and pharmacological means of activating these pathways to promote cell survival and plasticity will likely lead to novel approaches for promoting optimal health.

Acknowledgements This work was supported by the Intramural Research Program of the National Institute on Aging.

References

An JH, Blackwell TK (2003) SKN1 links *C. elegans* mesendodermal specification to a conserved oxidative stress response. Genes Dev 17: 1882–1893.

Anckar J, Hietakangas V, Denessiouk K et al. (2006) Inhibition of DNA binding by differential sumoylation of heat shock factors. Mol Cell Biol 26: 955–964.

Alessio HM, Goldfarb AH, Cutler RG (1988) MDA content in fast and slow-twitch skeletal muscle with intensity of exercise in a rat. Am J Physiol 255: C874–C877.

Atherton PJ, Higginson JM, Singh J et al. (2004) Concentration of signal transduction protein exercise and insulin responses in rat extensor digitorum longus and soleus muscles. Mol Cell Biochem 261: 111–116.

Baeuerle PA (1991) The inducible transcription factor NF-κB: regulation by distinct protein subunits. Biochem Biophys Acta 1072: 63–80.

Bahn YS, Xue C, Idnum A et al. (2007) Sensing the environment: lesson from fungi. Nat Rev Microbiol 5: 57–69.

Biamonti G (2004) Nuclear stress bodies: a heterochromatin affair?. Nat Rev Mol Cell Biol 5: 493–498.

Biggs WH, Meisenhelder J, Hunter T et al. (1999) Protein kinase B/AKT-mediated phosphorylation promotes nuclear exclusion of the winged helix transcription factor FKHR1. Proc Natl Acad Sci USA 96: 7421–7426.

Bishop NA, Guarente L (2007a) Genetic links between diet and lifespan: shared mechanisms from yeast to humans. Nat Rev Genet 8: 835–844.

Bishop NA, Guarente L (2007b) Two neurons mediate diet-restriction-induced longevity in *C elegans*. Nature 447: 545–549.

Bluher M, Khan BB, Kahn CR (2003) Extended longevity in mice lacking the insulin receptor in adipose tissue. Science 299: 572–574.

Boura E, Silhan J, Herman P et al. (2007) Both the N-terminal loop and wing W2 of the forkhead domain of transcription factor FoxO4 are important for DNA binding. J Biol Chem 282: 8265–8275.

Brunet A, Bonni A, Zigmond MJ et al. (1999) AKT promotes cell survival by phosphorylating and inhibiting a forkhead transcription factor. Cell 96: 857–868.

Brunet A, Kanai F, Stehn J et al. (2002) 14-3-3 transits to the nucleus and participates in dynamic nucleocytoplasmic transport. J Cell Biol 156: 817–828.

Brunet A, Sweeney LB, Sturgill JF et al. (2004) Stress-dependent regulation of FOXO transcription factors by the SIRT1 deacetylase. Science 303: 2011–2015.

Burgering BM, Kops GJ (2002) Cell cycle and death control: long live forkheads. Trends Biochem Sci 27: 352–360.

Cai D, Frantz JD, Tawa NE et al. (2004) IKKbeta/NF-kappaB activation causes severe muscle wasting in mice. Cell 119: 285–298.

Cameron AR, Anton S, Melville L et al. (2008) Black tea polyphenols mimic insulin/insulin-like growth factor-1 signaling to the longevity factor FOXO1a. Aging Cell 7: 69–77.

Chan K, Han XD, Kan YW (2001) An important function of nrf2 in combating oxidative stress: detoxification of acetaminophen. Proc Natl Acad Sci USA 98: 4611–4616.

Cho HY, Jedlicka AE, Reddy SP et al. (2002) Role of NRF2 in protection against hyperoxic lung injury in mice. Am J Respir Cell Mol Biol 26: 175–182.

Cho HY, Reddy SP, Debiase A et al. (2005) Gene expression profiling of NRF2-mediated protection against oxidative injury. Free Radic Biol Med 38: 325–343.

Chu B, Zhong R, Soncin F et al. (1998) Transcriptional activity of heat shock factor 1 at 37C is repressed through phosphorylation of two distinct residues by glycogen synthase kinase 3 and protein kinase Cα and Cζ. J Biol Chem 273: 18640–18646.

Clark KL, Halay ED, Lai E et al. (1993) Co-crystal structure of the HNF-3/fork head DNA recognition motif resembles histone H5. Nature 364: 412–420.

Clos J, Westwood JT, Becker PB et al. (1990) Molecular cloning and expression of a hexameric *Drosophila* heat shock factor subject to negative regulation. Cell 63: 1085–1097.

Cotman CW, Berchtold NC, Christie LA (2007) Exercise builds brain health: key roles of growth factor cascades and inflammation. Trends Neurosci 30: 464–472.

Daitoku H, Hatta M, Matsuzaki H et al. (2004) Silent information regulator 2 potentiates FoXo1-mediated transcription through its deacetylase activity. Proc Natl Acad Sci USA 101: 10042–10047.

Davies KJ, Quintanilha AT, Brooks GA et al. (1982) Free radicals and tissue damage produced by exercise. Biochem Biophys Res Commun 107: 1198–1205.

Durham WJ, Li YP, Genken E et al. (2004) Fatiguing exercise reduces DNA binding activity of NF-κB in skeletal muscle nuclei. J Appl Physiol 97: 1740–1745.

Denegri M, Moralli D, Rocchi M et al. (2002) Human chromosomes 9, 12, and 15 contain the nucleation sites of stress-induced nuclear bodies. Mol Biol Cell 13: 2069–2079.

Dinkova-Kostova AT, Holtzclaw WD, Cole RN et al. (2002) Direct evidence that sulphydryl groups of Keap1 are the sensors regulating induction of phase 2 enzymes that protect against carcinogens and oxidants. Proc Natl Acad Sci USA 99: 11908–11913.

Dowell P, Otto TC, Adi S et al. (2003) Convergence of peroxisome proliferator-activated receptor gamma and FOXO1 signaling pathways. J Biol Chem 278: 45485–45491.

Essers MA, Weijzen S, de Vries-Smits AM et al. (2004) FOXO transcription factor activation by oxidative stress mediated by small GTPase ral and JNK. EMBO J 23: 4802–4812.

Essers MA, de Vries-Smits LM, Barker N et al. (2005) Functional interaction between beta-catenin and FOXO in oxidative stress signaling. Science 308: 1181–1184.

Fujimoto M, Izu H, Seki K et al. (2004) HSF4 is required for normal cell growth and differentiation during mouse lens development. EMBO J 23: 4297–4306.

Furuyama T, Nakazawa T, Nakano I et al. (2000) Identification of the differential distribution patterns of mRNA and consensus binding sequences for mouse DAF-16 homologues. Biochem J 349: 629–634.

Giannakou ME, Goss M, Partridge L (2008) Role of dFOXO in lifespan extension by dietary restriction in *Drosophila melanogaster*: not required, but its activity modulates the response. Aging Cell 7: 187–198.

Gilley J, Coffer PJ, Ham J (2003) FOXO transcription factors directly activate Bim gene expression and promote apoptosis in sympathetic neurons. J Cell Biol 162: 613–622.

Gilmore TD (2008) www.nfkb.org.

Glauser DA, Schlegel W (2007) The emerging role of FOXO transcription factors in pancreatic β cells. J Endocrinol 193: 195–207.

Ghosh S, Karin M (2002) Missing pieces in the NF-kappaB puzzle. Cell 109: S81–S96.

Gomez-Cabrera MC, Borras C, Pallardó FV et al. (2005) Decreasing xanthine oxidase mediated oxidative stress prevents useful cellular adaptations to exercise in rats. J Physiol 567: 113–120.

Gomez-Cabrera MC, Martinez A, Santangelo G et al. (2006) Oxidative stress in marathon runners: interest of antioxidant supplementation. Br J Nutr 96: S31–S33.

Green M, Schultz TJ, Sullivan EK et al. (1995) A heat shock–responsive domain of human HSF1 that regulates transcription activation domain function. Mol Cell Biol 15: 3354–3362.

Greer EL, Brunet A (2005) FOXO transcription factors at the interface between longevity and tumor suppression. Oncogene 24: 7410–7425.

Gross SR, Kinzy TG (2005) Translation elongation factor 1a is essential for regulation of the actin cytoskeleton and cell morphology. Nat Struct Mol Biol 12: 772–778.

Gross DN, Van den Heuvel AP, Birnbaum MJ (2008) The role of FOXO in the regulation of metabolism. Oncogene 27: 2320–2336.

Guettouche T, Boellmann F, Lane WS et al. (2005) Analysis of phosphorylation of human heat shock factor 1 in cells experiencing stress. BMC Biochem 6: 4.

Han JM, Lee YJ, Lee SY et al. (2007) Protective effect of sulforaphane against dopaminergic cell death. J Pharmacol Exp Ther 321: 249–256.

Hansen M, Chandra A, Mitic LL et al. (2008) A role for autophagy in the extension of lifespan by dietary restriction in *C. elegans*. PLoS Genet 4: e24.

Hashimoto N, Kido Y, Uchida T et al. (2006) Ablation of PDK1 in pancreatic beta cells induces diabetes as a result of loss of beta cell mass. Nat Genet 38: 589–593.

Haskell WL, Lee IM, Pate RR et al. (2007) Physical activity and public health: updated recommendation for adults from the American College of Sports Medicine and the American Heart Association. Med Sci Sports Exerc 39: 1423–1434.

Hayden MS, Ghosh S (2008) Shared principles in NF-κB Signaling. Cell 132: 344–362.

He CH, Gong P, Hu B et al. (2001) Identification of activating transcription factor 4 (ATF4) as an Nrf2-interacting protein. Implication for heme oxygenase-1 gene regulation. J Biol Chem 276: 20858–20865.

He H, Soncin F, Grammatikakis N et al. (2003) Elevated expression of heat shock factor 2A stimulates HSF1-induced transcription during stress. J Biol Chem 278: 35465–35475.

Hietakangas V, Ahlskog JK, Jakobsson AM et al. (2003) Phosphorylation of serine 303 is a prerequisite for the stress-inducible SUMO modification of heat shock factor 1. Mol Cell Biol 23: 2953–2968.

Hsieh TC, Lu X, Wang Z et al. (2006) Induction of quinone reductase NQO1 by resveratrol in human K562 cells involves the antioxidant response element ARE and is accompanied by nuclear translocation of transcription factor Nrf2. Med Chem 3: 275–285.

Ho RC, Hirshman MF, Li Y et al. (2005) Regulation of IkappaB kinase and NF-kappaB in contracting adult rat skeletal muscle. Am J Physiol Cell Physiol 289: C794–C801.

Hoffmann A, Natoli G, Ghosh G (2006) Transcriptional regulation via the NF-kappaB signaling module. Oncogene 25: 6706–6716.

Hollander J, Bejma J, Ookawara T et al. (2000) Superoxide dismutase gene expression in skeletal muscle: fiber-specific effect of age. Mech Ageing Dev 116: 33–45.

Hollander J, Fiebig R, Ookawara T et al. (2001) Superoxide dismutase gene expression is activated by a single bout of exercise. Pflug Arch (Eur J Physiol) 442: 426–434.

Holmberg CI, Hietakangas V, Mikhailov A et al. (2001) Phosphorylation of serine 230 promotes inducible transcriptional activity of heat shock factor 1. EMBO J 20: 3800–3810.

Holzenberg M, Dupont J, Ducos B et al. (2003) IGF-1 receptor regulates lifespan and oxidative stress resistance in mice. Nature 421: 182–187.

Hong Y, Rogers R, Matunis MJ et al. (2001) Regulation of heat shock factor 1 by stress-induced SUMO-1 modification. J Biol Chem 276: 40263–40267.

Hunter RB, Kandarian SC (2004) Disruption of either the Nfkb1 or the Bcl3 gene inhibits skeletal muscle atrophy. J Clin Invest 114: 1504–1511.

Itoh K, Chiba T, Takahashi S, Ishii T et al. (1997) An Nrf2/small Maf heterodimer mediates the induction of phase II detoxifying enzyme genes through antioxidant response elements. Biochem Biophys Res Commun 236: 313–322.

Jeong WS, Keum YS, Chen C et al. (2005) Differential expression and stability of endogenous nuclear factor E2–related factor 2 (Nrf2) by natural chemopreventive compounds in HepG2 human hepatoma cells. J Biochem Mol Biol 38: 167–176.

Jolly C, Morimoto RI, Robert-Nicoud M et al. (1997) HSF1 transcription factor concentrates in nuclear foci during heat shock: relationship with transcription sites. J Cell Sci 110: 2935–2941.

Jolly C, Usson Y, Morimoto RI (1999) Rapid and reversible relocalization of heat shock factor 1 within seconds to nuclear stress granules. Proc Natl Acad Sci USA 96: 6769–6774.

Jolly C, Konecny L, Grady DL et al. (2002) In vivo binding of active heat shock transcription factor 1 to human chromosome 9 heterochromatin during stress. J Cell Biol 156: 775–781.

Jolly C, Metz A, Govin J et al. (2004) Stress-induced transcription of satellite III repeats. J Cell Biol 164: 25–33.

Kang ES, Woo IS, Kim HJ et al. (2007) Up-regulation of aldose reductase expression mediated by phosphatidylinositol 3-kinase/AKT and Nrf2 is involved in the protective effect of curcumin against oxidative damage. Free Radic Biol Med 43: 535–545.

Katob Y, Itoh K, Yoshida E et al. (2001) Two domains of Nrf2 cooperatively bind CBP, a CREB binding protein, and synergistically activate transcription. Genes Cells 6: 857–868.

Kensler TW, Wakabayashi N, Biswal S (2007) Cell survival responses to environmental stresses via the Keap1-Nrf2-ARE pathway. Annu Rev Pharmacol Toxicol 47: 89–116.

Kim SA, Yoon JH, Lee SH et al. (2005) Polo-like kinase phosphorylates heat shock transcription factor 1 and mediates its nuclear translocation during heat stress. J Biol Chem 280: 12653–12657.

Kirkwood TB, Shanley DP (2005) Food restriction, evolution and ageing. Mech Ageing Dev 126: 1011–1016.

Kitamura T, Nakae J, Kitamura Y et al. (2002) The forkhead transcription factor FoXo1 links insulin signaling to Pdx1 regulation of pancreatic beta cell growth. J Clin Invest 110: 1839–1847.

Kline MP, Morimoto RI (1997) Repression of the heat shock factor 1 transcriptional activation domain is modulated by constitutive phosphorylation. Mol Cell Biol 17: 2107–2115.

Knauf U, Newton EM, Kyriakis J et al. (1996) Repression of heat shock factor 1 activity at control temperature by phosphorylation. Genes Dev 10: 2782–2793.

Kobayashi A, Kang MI, Okawa H et al. (2004) Oxidative stress sensor Keap1 functions as an adaptor for Cul3-based E3 ligase to regulate proteasomal degradation of Nrf2. Mol Cell Biol 24: 7130–7139.

Kobayashi M, Yamamoto M (2005) Molecular mechanisms activating the Nrf2-Keap1 pathway of antioxidant gene regulation. Antioxid Redox Signal 7: 385–394.

Kong L, Tanito M, Huang Z, Li F et al. (2007) Delay of photoreceptor degeneration in tubby mouse by sulforaphane. J Neurochem 101: 1041–1052.

Kops GJ, Dansen TB, Polderman PE et al. (2002) Forkhead transcription factor FOXO3a protects quiescent cells from oxidative stress. Nature 419: 316–321.

Kraft AD, Johnson DA, Johnson JA (2004) Nuclear factor E2–related factor 2–dependent antioxidant response element activation by *tert*-butylhydroquinone and sulforaphane occurring preferentially in astrocytes conditions neurons against oxidative insult. J Neurosci 24: 1101–1112.

Kramer HF, Goodyear LJ (2007) Exercise, MAPK, and NF-κB signaling in skeletal muscle. J Appl Physiol 103: 388–395.

Kregel KC (2002) Heat shock proteins: modifying factors in physiological stress responses and acquired thermotolerance. J Appl Physiol 92: 2177–2186.

Kultz D (2005) Molecular and evolutionary basis of the cellular stress response. Annu Rev Physiol 67: 225–257.

Kwak MK, Wakabayashi N, Itoh K et al. (2003) Modulation of gene expression by cancer chemopreventive dithiolethiones through the Keap1-Nrf2 pathway. Identification of novel gene clusters for cell survival. J Biol Chem 278: 8135–8145.

Lee JM, Calkins MJ, Chan K et al. (2003a) Identification of the NF-E2–related factor-2–dependent genes conferring protection against oxidative stress in primary cortical astrocytes using oligonucleotide microarray analysis. J Biol Chem 278: 12029–12038.

Lee JM, Shih AY, Murphy TH et al. (2003b) NF-E2–related factor-2 mediates neuroprotection against mitochondrial complex I inhibitors and increased concentrations of intracellular calcium in primary cortical neurons. J Biol Chem 278: 37948–37956.

Leeuwenburgh C, Hollander J, Leichtweis M et al. (1997) Adaptation of glutathione antioxidant system to endurance training are tissue and muscle fiber specific. Am J Physiol 272: R363–R369.

Lehtinen MK, Yuan Z, Boaq PR et al. (2006) A conserved MST-FOXO signaling pathway mediates oxidative-stress responses and extends lifespan. Cell 125: 987–1001.

Libina N, Berman JR, Kenyon C (2003) Tissue-specific activities of *C. elegans* DAF-16 in the regulation of lifespan. Cell 115: 489–502.

Lindquist S (1986) The heat shock response. Annu Rev Biochem 55: 1151–1191.

Lindquist S, Craig EA (1988) The heat shock proteins. Annu Rev Genet 22: 631–677.

Martin D, Rojo AI, Salinas M et al. (2004) Regulation of heme oxygenase-1 expression through the phosphatidylinositol 3-kinase/AKT pathway and the Nrf2 transcription factor in response to the antioxidant phytochemical carnosol. J Biol Chem 279: 8919–8929.

McArdle A, Vasilaki A, Jackson M (2002) Exercise and skeletal muscle ageing: cellular and molecular mechanisms. Ageing Res Dev 1: 79–93.

Min KJ, Yamamoto R, Buch S et al. (2008) *Drosophila* lifespan control by dietary restriction independent of insulin-like signaling. Aging Cell 7: 199–206.
Minois N (2000) Longevity and aging: beneficial effects of exposure to mild stress. Biogerontology 1: 15–29.
Moi P, Chan K, Asunis I et al. (1994) Isolation of NF-E2–related factor 2 (Nrf2), a NF-E2-like basic leucine zipper transcriptional activator that binds to the tandem NF-E2/AP1 repeat of the β-globin locus control region. Proc Natl Acad Sci USA 91: 9926–9930.
Motta MC, Divecha N, Lemieux M et al. (2004) Mammalian SIRT1 represses forkhead transcription factors. Cell 116: 551–563.
Murray JI, Whitfield ML, Trinklein ND et al. (2004) Diverse and specific gene expression responses to stresses in cultured human cells. Mol Biol Cell 15: 2361–2374.
Muto A, Tashiro S, Tsuchiya H et al. (2002) Activation of Maf/AP-1 repressor Bach2 by oxidative stress promotes apoptosis and its interaction with promyelocytic leukemia nuclear bodies. J Biol Chem 277: 20724–20733.
Myzak MC, Dashwood RH (2006) Chemoprotection by sulforaphane: keep one eye beyond Keap1. Cancer Lett 233: 208–218.
Nakae J, Biggs WH 3rd, Kitamura T et al. (2002) Regulation of insulin action and pancreatic beta-cell function by mutated alleles of the gene encoding forkhead transcription factor FoXo1. Nat Genet 32: 245–253.
Nakai A, Tanabe M, Kawazoe Y et al. (1997) HSF4, a new member of the human heat shock factor family which lacks properties of a transcriptional activator. Mol Cell Biol 17: 469–481.
Negrutskii BS, El'skaya AV (1998) Eukaryotic translation elongation factor 1 alpha: structure, expression, functions, and possible role in aminoacyl-t-RNA channeling. Prog Nucleic Acid Res Mol Biol 60: 47–78.
Nguyen T, Yang CS, Pickett CB (2004) The pathways and molecular mechanisms regulating Nrf2 activation in response to chemical stress. Free Radic Biol Med 37: 433–441.
Ostling P, Biork JK, Ross-Mattjus P et al. (2007) Heat shock factor 2 (HSF2) contributes to inducible expression of hsp genes through interplay with HSF1. J Biol Chem 282: 7077–7086.
Panniers R (1994) Translational control during heat shock. Biochemie 76: 737–747.
Pearson KJ, Lewis KN, Price NL et al. (2008) Nrf2 mediates cancer protection but not prolongevity induced by caloric restriction. Proc Natl Acad Sci USA 105: 2325–2330.
Pirkalla L, Nykanen P, Sistonen L (2001) Roles of the heat shock transcription factors in regulation of the heat shock response and beyond. FASEB J 15: 1118–1131.
Plas DR, Thompson CB (2003) AKT activation promotes degradation of tuberin and FOXO3a via the proteasome. J Biol Chem 278: 12361–12366.
Rabindran SK, Haroun RI, Clos J et al. (1993) Regulation of heat shock factor trimer formation: role of a conserved leucine zipper. Science 259: 230–234.
Radak Z, Kaneko T, Tahara S et al. (1999) The effect of exercise training on oxidative damage of lipids, proteins and DNA in rat skeletal muscle: evidence for beneficial outcomes. Free Radic Biol Med 27: 69–74.
Radak Z, Kaneko T, Tahara H et al. (2001) Regular exercise improves cognitive function and decreases oxidative damage in rat brain. Neurochem Int 38: 17–23.
Radak Z, Chung HY, Goto S (2005) Exercise and hormesis: oxidative stress-related adaptation for successful aging. Biogerontology 6: 71–75.
Rallu M, Loones M, Lallemand Y et al. (1997) Function and regulation of heat shock factor 2 during mouse embryogenesis. Proc Natl Acad Sci USA 94: 2392–2397.
Ramos-Gomez M, Kwak MK, Dolan PM et al. (2001) Sensitivity to carcinogenesis is increased and chemoprotective efficacy of enzyme inducers is lost in nrf2 transcription factor-deficient mice. Proc Natl Acad Sci USA 98: 3410–3415.
Rattan SIS (2004) Aging intervention, prevention and therapy through hormesis. J Gerontol Biol Sci 59A: 705–709.
Rena G, Prescott AR, Guo S et al. (2001) Roles of the forkhead in rhabdomyosarcoma (FKHR) phosphorylation sites in regulating 14-3-3 binding, transactivation and nuclear targeting. Biochem J 354: 605–612.

Rena G, Woods YL, Prescott AR et al. (2002) Two novel phosphorylation sites on FKHR that are critical for its nuclear exclusion. EMBO J 21: 2263–2271.
Ritossa F (1962) A new puffing pattern induced by temperature shock and DNP in *Drosophila*. Experimentia 18: 571–573.
Rubiolo JA, Mithieux G, Vega FV (2008) Resveratrol protects primary rat hepatocytes against oxidative stress damage. Activation of the Nrf2 transcription factor and augmented activities of antioxidant enzymes. Eur J Pharmacol 591: 66–72.
Shamovsky I, Ivannikov M, Kandel ES et al. (2006) RNA-mediated response to heat shock in mammalian cells. Nature 440: 556–560.
Shamovsky I, Nudler E (2008) New insights into the mechanism of heat shock response activation. Cell Mol Life Sci 65: 855–861.
Shankar S, Chen Q, Srivastava RK (2008) Inhibition of PI3K/AKT and MEK/ERK pathways act synergistically to enhance antiangiogenic effects of EGCG through activation of FOXO transcription factor. J Mol Signal 20: 3–7.
Shen G, Jeong WS, Hu R, Kong AN (2005) Regulation of Nrf2, NF-kappaB, and AP-1 signaling pathways by chemopreventive agents. Antioxid Redox Signal 7: 1648–1663.
Senftlenben U, Cao Y, Xiao G et al. (2001) Activation by IKKα of a second evolutionary conserved, NF-κB signaling pathway. Science 293: 1495–1499.
Son TG, Camandola S, Mattson MP (2008) Hormetic dietary phytochemicals. Neuromolecular Med 10: 236–246.
Sun J, Hoshino H, Takaku K et al. (2002) Hemoprotein Bach1 regulates enhancer availability of heme oxygenase-1 gene. EMBO J 21: 5216–5224.
Surh YJ (2003) Cancer chemoprevention with dietary phytochemicals. Nat Rev Cancer 3: 768–780.
Tanabe M, Sasai N, Nagata K et al. (1999) The mammalian HSF4 gene generates both an activator and a repressor of heat shock genes by alternative slicing. J Biol Chem 274: 27845–27856.
Tanigawa S, Fujii M, Hou DX (2007) Action of Nrf2 and Keap1 in ARE-mediated NQO1 expression by quercetin. Free Radic Biol Med 42: 1690–1703.
Thimmulappa RK, Mai KH, Srisuma S et al. (2002) Identification of Nrf2-regulated genes induced by the chemopreventive agent sulforaphane by oligonucleotide microarray. Cancer Res 62: 5196–5203.
Tissembaum HA, Guarente L (2001) Increased dosage of Sir-2 gene extends lifespan in *Caenorhabditis elegans*. Nature 410: 227–230.
Tran H, Brunet A, Grenier JM et al. (2002) DNA repair pathway stimulated by the forkhead transcription factor FOXO3a through the Gadd45 protein. Science 296: 530–534.
Tsai WC, Bhattacharyya N, Han LY et al. (2003) Insulin inhibition of transcription stimulated by forkhead protein FOXO1 is not solely due to nuclear exclusion. Endocrinology 144: 5615–5622.
Tullet JMA, Hertweck M, An JH et al. (2008) Direct inhibition of the longevity-promoting factor SKN-1 by insulin-like signaling in *C. elegans*. Cell 132: 1025–1038.
van der Heide L, Smidt MP (2005) Regulation of FOXO activity by CBP/p300-mediated acetylation. Trends Biochem Sci 30: 81–86.
van der Host A, de Vries-Smits AM, Brenkman AB et al. (2006) FOXO4 transcriptional activity is regulated by monoubiquitination and USP7/HAUSP. Nat Cell Biol 10: 1064–1073.
van der Horst A, Burgering BM (2007) Stressing the role of FOXO proteins in lifespan and disease. Nat Rev Mol Cell Biol 8: 440–450.
Venugopal R, Jaiswal AK (1998) Nrf2 and Nrf1 in association with Jun proteins regulate antioxidant response element–mediated expression and coordinated induction of genes encoding detoxifying enzymes. Oncogene 17: 3145–3156.
Vogt PK, Jiang H, Aoki M (2005) Triple layer control: phosphorylation, acetylation and ubiquitination of FOXO proteins. Cell Cycle 4: 908–913.
Voellmy R (2004) On mechanisms that control heat shock transcription factor activity in metazoan cells. Cell Stress Chaperones 9: 122–133.

Wakabayashi N, Itoh K, Wakabayashi J et al. (2003) Keap1-null mutation leads to post-natal lethality due to constitutive Nrf2 activation. Nat Genet 35: 238–245.

Wang X, Grammatikakis N, Siganou A et al. (2003) Regulation of molecular chaperons gene transcription involves the serine phosphorylation, 14-3-3 epsilon binding, and cytoplasmatic sequestration of heat shock factor 1. Mol Cell Biol 23: 6013–6026.

Wang X, Grammatikakis N, Siganou A et al. (2004) Interactions between extracellular signal regulated protein kinase 1 (ERK1), 14-3-3 epsilon and heat shock factor 1 during stress. J Biol Chem 279: 49460–49469.

Wang MC, Bohmann D, Jasper H (2005) JNK extends life span and limits growth by antagonizing cellular and organism-wide responses to insulin signaling. Cell 121: 115–125.

Welch WJ, Suhan JP (1985) Morphological study of the mammalian stress response: characterization of changes in cytoplasmic organelles, cytoskeleton, and nucleoli, and appearance of intranuclear actin filaments in rat fibroblasts after heat-shock treatment. J Cell Biol 101: 1198–1211.

Westerheid SD, Morimoto RI (2005) Heat shock response modulators as therapeutic tools for disease of protein conformation. J Biol Chem 280: 33097–33100.

Wu C (1995) Heat shock transcription factors: structure and regulation. Annu Rev Cell Dev Biol 11: 441–469.

Wu CC, Hsu MC, Hsieh CW et al. (2006) Upregulation of heme oxygenase-1 by epigallocatechin-3-gallate via the phosphatidylinositol 3-kinase/AKT and ERK pathways. Life Sci 78: 2889–2897.

Xiao H, Perisic O, Lis JT (1991) Cooperative binding of *Drosophila* heat shock factor to arrays of a conserved 5 bp unit. Cell 64: 585–593.

Xiao X, Zuo X, Davis AA et al. (1999) HSF1 is required for extra-embryonic development, postnatal growth and protection during inflammatory responses in mice. EMBO J 18: 5943–5952.

Zhang DD, Hannink M (2003) Distinct cysteine residues in Keap1 are required for Keap1-dependent ubiquitination of Nrf2 and for stabilization of Nrf2 by chemopreventive agents and oxidative stress. Mol Cell Biol 23: 8137–8151.

Zhang DD (2006) Mechanistic studies of the Nrf2-Keap1 signaling pathway. Drug Metab Rev 38: 769–789.

Zou J, Guo Y, Guettouche T et al. (1998) Repression of heat shock factor HSF1 activation by HSP90 (HSP90-complex) that forms a stress sensitive complex with HSF1. Cell 94: 471–480.

The Devil Is in the Dose: Complexity of Receptor Systems and Responses

Wayne Chadwick and Stuart Maudsley

Abstract Through evolutionary history the primary mechanism by which the cells or tissues of most organisms sense their environment has been the heptahelical G protein–coupled receptor (GPCR). This prototypic receptive entity has its origins in the earliest forms of life and often comprises up to 5% of the genome of most unicellular and multicellular life forms. The GPCR system has adapted to perceive almost all forms of environmental entities, for example, photons, odorants, lipids, carbohydrates, peptides, and nucleic acids. The GPCR system has also likely adapted to the presence of exogenous compounds that may at some doses be deleterious but at lower levels may indeed possess beneficial actions. Therefore, with respect to the evolutionary pressure of diverse environments, it would be an extreme advantage for an organism to adapt multiple components of its primary receptive system to take advantage of any beneficial effects of agents present in harsh or damaging environments.

Keywords G protein–coupled receptor · Dose response · Allosteric · Conformation · Environment · Flavor · Receptive

Introduction

The branch of science known as pharmacology can be considered to have been started by Philip von Hohenheim (1493–1541). Although his original birth name may not be familiar, he is better known among pharmacologists and molecular biologists as "Paracelsus," the name he coined to liken himself to the Roman encyclopedist Celsus. In addition to being credited for naming the metal zinc, he is best known for his comments regarding the toxicity of substances to human beings. He

S. Maudsley (✉)
Receptor Pharmacology Unit, Laboratory of Neurosciences, National Institute on Aging, National Institutes of Health, Biomedical Research Center, Baltimore, MD 21224, USA
e-mail: maudsleyst@grc.nia.nih.gov

M.P. Mattson, E.J. Calabrese, *Hormesis*, DOI 10.1007/978-1-60761-495-1_5,

stated, "*Alle Ding sind Gift, und nichts ohn Gift; allein die Dosis macht, daß ein Ding kein Gift ist*," which can be literally translated as, "All things are poison and nothing is without poison, only the dose permits something not to be poisonous." This was the first strong statement of the concept of the dose dependence of effects. From this point, on the perception of the beneficial or noxious effects of compounds was changed. Anecdotally this was long known, as evidenced in the use of often-"toxic" agents in medicine, such as ergot alkaloids for parturition control. With respect to modern pharmacology, Paracelsus' words still hold true, given that essentially even the most beneficial compound can exert deleterious effects, usually at higher doses than prescribed therapeutically. An important concern, and one that we specifically address in this chapter, is the nature of these differential effects of compounds or therapeutics at various doses. Quite often the assumption is made that at high concentrations drug effects can be considered to be nonspecific yet are still related to the qualitative nature of the effects of lower doses. It is more parsimonious to consider that responses at the maximal end of classic sigmoidal log dose-response relationships are the result of composite actions of competing effects. In a similar manner recent hypotheses have begun to consider the nature of responses at the very lower end of the drug-response relationship. The theory known as "hormesis" in part attempts to delineate and support the benefits of ultralow doses of usually noxious or toxic agents. This theory suggests that biological systems can often respond to low doses with a beneficial response as this manageable insult then inculcates a long-lasting protective response from the organism to the insult in a similar manner to, say, active immunization paradigms. For any agent to exert an effect on a secondary entity (i.e., organism) some level of transmission through a "receptive" system is required. By far the best-studied "receptive" system in physiological settings is the rhodopsin-like (class I) G protein–coupled receptor (GPCR) (Fredriksson and Schiöth, 2005). Perhaps our understanding of hormetic mechanisms can be illuminated by considering the functioning of the GPCR superfamily, given that this group of proteins has been demonstrated to be the most pluripotent receptive structure in the genome of virtually every organism.

In this chapter we discuss how responses to drugs or other outside agents can exhibit complex effects by considering how receptive systems respond at multiple doses or degrees of input stimulation. We consider how, at different doses, the complexity of plasma membrane receptor systems may lend itself to the creation of seemingly divergent deleterious or beneficial reactions to xenobiotic or endogenous compounds.

Classic and Modern Dynamic GPCR Models

G protein–coupled receptors act at the most basic level as ligand-gated guanine nucleotide exchange factors for the heterotrimeric class of G proteins. Ligands that bind to the receptor and positively stimulate it to cause an exchange of GDP (guanosine diphosphate) for GTP (guanosine triphosphate) at the alpha subunit of the heterotrimeric G protein were classically considered agonists. Ligand molecules that could bind the receptor but could not positively induce this nucleotide exchange were denoted as antagonists. Early research verifying this simplistic model demonstrated that there were two distinct agonist affinity states of the receptor, and that the relative populations of the two receptor affinity states were determined by the intimate presence of GDP or GTP with the receptor. In the absence of GTP, receptors were primarily in the low-affinity, "inactive" state, with the high-agonist-affinity, "active" state being promoted by the presence of GTP (DeLean et al., 1980). The hypothetical models that described this tripartite molecular interaction—that is, ligand (A), receptor (R), and G protein (G)—were described collectively as ternary complex models. Further research, however, demonstrated the presence of additional states of the receptor intermediate between ligand binding and adoption of the "active" state (Samama et al,. 1993). In this model—the "extended ternary complex model"—based on mutation of adrenergic receptors, it was noted that in the absence of agonist ligand the receptor can adopt a high-affinity state that is then stabilized by the agonist on binding. It also was hypothesized that the agonist instead may choose between "inactive" or "active" states of the receptor that are transiently created by spontaneous receptor isomerization. It is largely regarded that both mechanisms exist in nature and are not mutually exclusive. Therefore in this model the common facet is that the GPCR exists in a spontaneous equilibrium between the inactive (R) and the active (R*) state. The efficacy of an agonist ligand in this model is determined by its ability to either select for the active state or to stabilize it compared to the inactive state. This model was able to accommodate experimental observations of positive agonism (ligand preference for R* over R), neutral antagonism (no preference of ligand between R* and R), and also inverse agonism (ligand preference of R over R*). The last of these agonistic properties was only revealed on the creation of receptor mutants that possessed signaling activity in the absence of traditional agonists. Although these models formed the basis of our understanding of many receptor systems, almost a decade of further research led to additions to these early theories. Two of the more important facets that required attention were the identification that receptor systems consisted of more protein molecules than described in the ternary and extended ternary complex models and the demonstration that multiple additional receptor states can exist [for review see Maudsley et al. (2005)]. Physical experimental findings therefore dictated that alterations to the classical receptor theory needed to be introduced (Kenakin, 2002; Kenakin, 2003). Therefore it is now implicit in many models of receptor activity that the receptive agent may exist in multiple receptive conformations that can be differentially stabilized by multiple extra protein–protein interactions. These added complexities not only need to form the basis for the eventual development of additional therapeutics (Maudsley et al., 2007b), but they may also provide mechanisms by which hormetic actions can be mediated. We therefore describe how these complexities arise in GPCR signaling and hypothesize how they might be exploited by agents that demonstrate hormetic tendencies.

Receptor System Complexities and Responses

Multiple G Protein Coupling

Receptor conformation, as we have described, can be strongly controlled by activity status. In addition, multiple conformations can also be recognized/stabilized by the individual nature of the ligand that interacts with the receptor (Pommier et al., 1999; Sagan et al., 1996; Maudsley et al., 1998; Maudsley et al., 2004). Therefore several distinct active states of the receptor seem to be promoted/stabilized by diverse ligands binding the receptor. At the most rudimentary level this diversity of functional status of the receptor is linked to the type of G protein the receptor can interact with. The specificity of G protein interactivity is likely to be controlled by the plasma membrane microenvironment in which the receptor is resident. Indeed GPCRs that often obligatorily couple to one type of G protein can be coerced, through interaction with intracellular scaffolding molecules, to interact with a different class of G proteins (Maudsley et al., 2004). From this aspect, if G protein coupling can be specified by cellular location or specific agonistic ligand, it is possible that for a single receptor "stimulation event" multiple response pathways may be activated in either a hierarchical or simultaneous manner. Thus with multiple ligand doses a vast range of potential divergent ligand-mediated responses may be seen. It has been proposed that variation in such a mechanism may account for seemingly paradoxical receptor–ligand interactions, that is, ligands possessing both pro- and antiapoptotic receptor-mediated actions (Rimoldi et al., 2003; Chadwick et al., 2008). Therefore, depending on the sequence of preferential G protein pathway stimulation, an agonistic agent may activate some pathways at low doses, a mixture of pathways at intermediate doses, and then another, select group of pathways at even higher doses (Fig. 1).

With specific respect to the ability to select for different G protein–coupled receptor populations, the original historical description of classical GPCR ligands often breaks down. Hence it has been shown that both antagonists (Maudsley et al., 2004) and inverse agonists (Zhang et al., 2000; Maudsley et al., 2000) can act in a similar "stimulatory" manner to classic agonists but just through different G protein–coupled mechanisms. For example, the serotonin 5-HT_{2C} receptor is able to activate different G proteins (Gq, Gαi/o, or Gα12/13), depending on the chemical nature of the ligand that is interacting with the receptor (Berg et al., 1998). Therefore the functional classification of the ligand is also contextually derived, as well as the responsivity of the receptor system. In the face of such pleiotropy of receptive systems and their ligands, it is perfectly feasible to expect often-complicated and seemingly paradoxical dose-response profiles. Therefore, environmentally derived

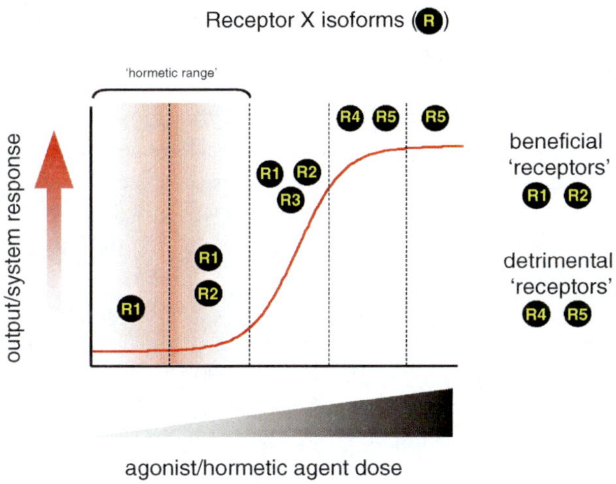

Fig. 1 Dose-dependent receptor isoform diversity: potential mechanism for hormetic activity. The rhodopsin-like GPCR, X, can naturally exist in multiple structural isoforms through accessory protein–controlled isomerization. Each of these isoforms (1–5) may be preferentially coupled to discrete signaling pathways and therefore control selective biological activities. The effect of accessory protein control on receptor isoform state also may affect the affinity/efficacy of a given ligand or hormetic agent for the receptor. At lower doses the ligand may only bind to and activate a small number of specific isoforms of X (R1, R2), which may then be coupled to physiologically beneficial signaling pathways. At higher and higher doses there is an alteration in the recruitment of receptor isoforms that then may not induce beneficial pathways but actually activate signaling pathways that are detrimental to the organism or cell (R4, R5)

agents that may in some way mimic endogenous GPCR ligands may create complicated dose-response relationships that could reproduce and underlie observed hormetic effects.

Allosteric Receptor Modulation

The molecular interaction between ligand and GPCR that results in activation of the receptor has been primarily considered to occur in the three-dimensional space in which the endogenous native ligand for that receptor binds reversibly with high affinity. For rhodopsin-like GPCRs this binding region, deep in the heptahelical core for small biogenic amines and catecholamines, and superficially with the extracellular loops for small neuropeptides, is termed the orthosteric biding site. Productive occupation of this site by an agonist that catalyzes GDP for GTP exchange at the associated G protein is sufficient for activation of the receptor. However, for most of the rhodopsin-like GPCRs there are additional sites, outside the orthosteric binding region, that allow additional ligands to undergo a high-affinity interaction with the receptor that can affect the signaling efficiency of the orthosteric ligand. These additional binding compounds are termed allosteric modulators (AMs). Most AM agents do not directly activate the GPCR but serve to either increase or decrease the efficiency by which the orthosteric ligand can activate the receptor (Christopoulos and Kenakin, 2002; Christopoulos et al., 2004; May et al., 2007). These allosteric mechanisms have been demonstrated in a huge variety of rhodopsin-like receptors. For example, the muscarinic, adenosine, and chemokine CCR5 receptors possess small allosteric ligand-binding sites that are structurally and spatially distinct from the receptor orthosteric sites (Christopoulos and Kenakin, 2002; Lazareno, 2004; Soudijn, 2004; Birdsall and Lazareno, 2005; Watson, 2005). A particularly well-studied example of this novel complexity in GPCR signaling is the β1-adrenoceptor. This rhodopsin-like GPCR has been shown to possess more than one agonist-binding site, each with unique pharmacological properties. These two ligand-binding sites appear to be functionally independent of one another (Granneman, 2001; Moolenaar, 2003; Arch, 2004). The presence of this second binding site was demonstrated using CGP 12177, a high-affinity binding antagonist of the β1-adrenoceptor. This antagonist displayed inhibition of isoproterenol-mediated β1-adrenoceptor activation at low concentrations but served as an agonist at higher concentrations (Pak and Fishman, 1996; Konkar, 2000; Lowe, 2002). The classic β-adrenoceptor orthosteric agonist isoproterenol is selectively and competitively inhibited by the orthosteric antagonist CGPA. However, at concentrations of CGPA that can effectively inhibit isoproterenol-induced receptor activation, specific doses of CGP 12177 were still able to stimulate the adrenoceptor (Baker, 2005). To confirm that indeed CGP 12177 was binding at a distinct site to the orthosteric agents (isoproterenol or CGPA), its dissociation constant from the receptor was shown to be significantly different from that of 17 other orthosteric ligands (Baker, 2005). Therefore it is likely that even without disrupting the binding of the orthosteric endogenous ligand to the receptor, a hormetic agent could still exert a functional action at the receptor by subtly altering how the orthosteric ligand can exert its effects. The hormetic agent could possibly affect the affinity or dissociation constant of the orthosteric ligand for its primary binding site or alter the ability of the orthosteric ligand to stabilize/induce specific active conformations of the receptor.

Allosteric modulation may also mediate some hormetic actions by inducing a shift in receptor–ligand stoichiometric homeostasis. An allosteric inhibitor could lead to the partial or total decrease in receptor efficacy irrespective of the amount of available ligand. This would thereby prompt the cell to upregulate the receptor to facilitate the desired response, leading to a massive increase in receptor expression that would leave the cell at either an advantage or a disadvantage once the allosteric inhibitor is removed. The organism may also compensate for the decreased signaling by increasing the availability of endogenous ligand, which again may be beneficial or harmful, depending on the system that is perturbed. The reverse would of course apply in the case of allosteric activators. Allosteric activators could possibly affect the conformation of the receptor's carboxyl terminus, thereby influencing the interaction with intracellular binding proteins that may stabilize the receptor's expression at the plasma membrane or lead to its increased internalization.

AM agents interact with regions of the receptor that are not conserved for cognate ligand binding, and these sites frequently show a high degree of three-dimensional molecular diversity, even among multiple receptors that share a common cognate ligand. Such a functional niche in GPCR signaling seems an ideal mechanism by which either xenobiotic chemicals or endogenous compounds from different species could affect the physiology/pathophysiology of humans. The majority of molecules identified so far that act as positive allosteric modulators (enhancing orthosteric ligand function) or negative allosteric modulators (inhibiting orthosteric ligand function) are small inorganic compounds with molecular masses of less than 500 to 600 atomic mass units (Birdsall et al., 1983; Waelbroeck 1994; Marlo et al., 2008). Such small agents therefore could be easily produced as a by-product of chemical/physical synthetic processes and also be easily absorbed by inhalation or water or skin contact. These relatively small inorganic compounds would also likely have a reasonably good bioavailability if they could avoid an initial "first-pass" degradation by the liver. Therefore, with respect to hormetic mechanisms, it is highly likely that allosteric GPCR modulation could be one of their loci of activity.

Receptor Desensitization

The ability of rhodopsin-like GPCRs to rapidly attenuate their signal productivity after initial agonist stimulation was previously considered to be a simplistic inhibitory mechanism. However, research over the last decade has demonstrated that interaction of the GPCR with the molecular machinery controlling receptor desensitization actually modulates signaling direction more than by simple termination (Daaka et al., 1997; Maudsley et al., 1998; Luttrell et al., 1999; Stout and Clarke 2002, Sneddon et al., 2003; Kohout et al., 2004). In the classic two-state GPCR dynamic model the degree of signal intensity should be largely proportional to the extent of desensitization; however, many examples of a disparity exist, one of the best studied being the μ-opioid receptor. Endogenous opioid peptide activation of the μ-opioid receptor induces a typical reflexive G protein–coupled receptor kinase–mediated phosphorylation and desensitization of the receptor. However, the xenobiotic opioid receptor agonists methadone and buprenorphine induce far more receptor phosphorylation and desensitization than for a similar dose of endogenous opiate (Yu et al., 1996). The plant alkaloid morphine, a non-peptidergic agonist of the μ-opioid receptor, however, can achieve the converse, that is, it can induce intense receptor activation but then does not induce a reflexive desensitization and internalization (Keith et al., 1996). To underline the truly huge variety of potential signaling/dynamic outcomes of ligand interaction with GPCRs, the converse to the action of morphine on μ-opioid receptors has been shown, that is, the ability of cholecystokinin, angiotensin, and serotonin receptor antagonists to induce rapid and profound internalization while actually blocking functional G protein activation (Roettger et al., 1997; Holloway et al., 2002; Peroutka and Snyder 1980). In addition to these differential ligand-mediated desensitization activities, disparities in the actual recovery rate from desensitization have also been described for different ligands acting at the same receptor (van Hooft and Vijverberg, 1996). These multiple and profound distinctions induced on receptor activity by structurally divergent, often nonnatural ligands potentially demonstrate the presence of specific subpopulations of GPCRs linked either to specific signal transduction machinery or decompartmentalized away from desensitization/internalization machinery. Thus, with respect to the presence of external (potentially hormetic) agents that may exist in the environment, if these can functionally interact with GPCRs, they may also exert a specific and discrete effect on GPCR desensitization and internalization. Such activities may again subtly alter GPCR-mediated signaling paradigms in cells and tissues that could result in either deleterious or beneficial physiological actions, depending upon the degree and mode of GPCR signaling/sensitivity modulation.

Receptor Dimerization

The heptahelical core of GPCRs was primarily considered to be the unique receptive entity with respect to stimulatory GPCR ligands; however, multiple sources of experimental data have demonstrated that GPCRs can form functional dimeric or multimeric structures (Angers et al., 2002; Urban et al., 2007). The formation of multimeric GPCR structures is thought to facilitate the creation of multiple levels of additional signaling "texture." The dimeric and multimeric structures themselves are not thought to be maintained by any form of covalent intermolecular bonds but are potentially reinforced by high-density lipid microdomains or through secondary interactions with intracellular scaffolding proteins. Guo et al. (2005) and Liang et al. (2003) proposed models for the positioning of receptors juxtaposed to each other in the plasma membrane. The orientation of the dimerizing receptor interfaces in the complex was proposed to control the resultant activation of the multireceptor unit. Functional dimerization in some receptor cases is even vital for their plasma membrane expression; for example, $GABA_B$R1 and R2 receptors must form a dimeric structure in the endoplasmic reticulum to facilitate their efficient trafficking to the plasma membrane (Jones et al., 1998). The combination of divergent receptor types into dimeric structures has been shown to affect their binding affinity for certain agonists (Franco et al., 2000); to negatively control each others' signaling activity (Jordan and Devi, 1999); to alter the quality of the resultant output receptor dimer signaling (George et al., 2000); and even to alter the eventual desensitization and internalization of the receptors associated with each other (Jordan and Devi, 1999). Not only can receptor dimerization affect gross receptor function, but also through a process described as "domain swapping" the ligand-binding sites can be qualitatively disrupted. "Domain swapping" refers to the concept that for small peptidergic ligands of rhodopsin-like GPCRs it is possible for them to directly interact with two receptor heptahelical cores at one time (given that the receptors are within sufficient proximity of each other). Thus the actual binding site itself may only be "presented" during certain times of dimerization, and also its occupation may impart entirely different agonistic or antagonistic properties to the bound ligand. It is therefore reasonable to suggest that xenobiotic or environmentally derived agents that could promote hormetic effects may additionally alter the dimerization of GPCRs and then cause multiple and complex changes to ligand-response profiles in both a qualitative and a quantitative manner.

GPCRs and Receptorsome Structures

One of the most exciting fields of GPCR research in recent years has been concerned with the elucidation of the surprising number of additional functional proteins that are required for the full range of GPCR activities (Brady and Limbird, 2002; Bockaert et al., 2003; Maudsley, et al. 2005). These proteins take the form of either transmembrane proteins or intracellular scaffolding/signaling proteins that possess the capacity to maintain an intimate contact with the plasma membrane. The range and variety of proteins that show a functional effect on receptor pharmacology is constantly expanding and now comprises adapter proteins such as β-arrestin (Luttrell et al., 1999), spinophilin (Wang et al., 2004), and RAMP (Sexton et al., 2001); scaffolding proteins such as NHERF (Maudsley et al., 2000a) and Pyk2 (Maudsley et al., 2007a); tyrosine kinases, such as Jak2 (Ali et al., 1997); the epidermal growth factor receptor (Maudsley et al., 2000); and endocytic proteins such as SH3-p4 (Tang et al., 1999) and adaptin (Laporte et al., 1999). An important distinction between these binding factors and the classic GPCR interactors, that is, G proteins, is the stability and longevity of their interaction. Through techniques such as co-immunoprecipitation, reliable and stable interaction of these receptor-modifying proteins with GPCRs can de demonstrated. In contrast, the interaction with G proteins is relatively transient in nature. Thus, for the majority of the lifetime of a GPCR it is in stable connection with a multitude of proteins, a superstructure that has been dubbed the "receptorsome." This higher-order "receptorsome" structure facilitates the stability of the GPCR at the membrane, localization to specific microdomains, and ease of coupling to downstream signaling cascades (Maudsley et al., 2005). This pre-organization of receptors into these structures poses problems with respect to classic theories of receptor–effector coupling, that is, rapid conversion from activation of one transduction pathway to the next becomes increasingly difficult with this elevation in receptor stability and complexity. Therefore it is more likely that GPCRs may be pre-organized into essentially "hard-wired" signaling structures and that the variety of receptor–effector coupling is generated by the creation of multiple divergent receptorsome structures (Maudsley et al., 2005). This pre-organization may also impose crucial spatial resolution and signaling compartmentalization on signaling systems required for rapid and high-fidelity physiological regulation. These different structural and functional receptorsome isoforms, or "flavors" of the same receptor, may demonstrate distinct signal–effector coupling, as well as altered ligand interaction (Maudsley et al., 2004; Chadwick et al., 2008). These multiple receptor substates may also underlie the ability of exogenous or endogenous hormetic compounds to elicit complex dose-response relationships, that is, at the lowest doses one specific type of receptor "flavor" may be stimulated alone, producing, say, a protective hormetic effect through a unique signaling pathway, whereas at higher concentrations, more and more different "flavors" of receptors are activated, generating responses distinct from that of the very low doses of the hormetic agent. It is also possible that at even higher concentrations a single receptor "flavor" may only be stimulated. In addition, there may be certain receptor "flavors" that may be rendered relatively insensitive to their original endogenous ligand but through random mutagenesis or evolutionary pressure possess a higher affinity/efficacy of transduction for an exogenous hormetic compound.

Conclusions

We have attempted to address the issue of how potentially hormetic actions of xenobiotic agents can be interpreted through an understanding of the complexity of G protein–coupled receptor signaling systems. Given that these sensitive and nearly ubiquitous sensory systems have been conserved from the onset of biological life, both in protists and eukaryotes, it is likely that they have evolved mechanisms to detect and/or utilize agents present in the environment. To ensure species survival, specific organisms may have been able to utilize environmental agents, poisonous or beneficial, to aid their biochemical or physiological status (Fig. 2). Such an advantage may have secured the capacity of early life forms to take a foothold in the harsh primordial terrestrial or oceanic environments. It is therefore also likely that successful species possess relatively plastic "receptive" systems (i.e., GPCRs) to ensure that this potential advantageous sensitivity is maintained to deal with potential future changes in the environment. Considering this postulate, it is therefore appropriate that there is a high percentage of the genome of simple and complex organisms that is given over to the encoding of GPCR genes (Fredriksson and Schiöth, 2005). It is likely that organisms that possess the greatest flexibility in their GPCR systems may be the most successful ones in a given environment because they possess the capacity to utilize the greatest variety of environmental agents for potentially hormetic effects. It is interesting to note that organisms that come into close contact with multiple environmental agents and do not possess complex immune or defense mechanisms, for example, nematode worms, have a much greater percentage of their genome given over to GPCRs (5.5%) (Takeda et al., 2002) than do more complex organisms such as humans (approximately 1.5%) that possess strong defensive barriers such as multiple skin layers and active immune systems. Therefore in part the evolution of the GPCR system may be tightly connected with the ability of organisms to first sustain life and reproduction in the primeval Earth environment and then to prosper in future uncertain environments.

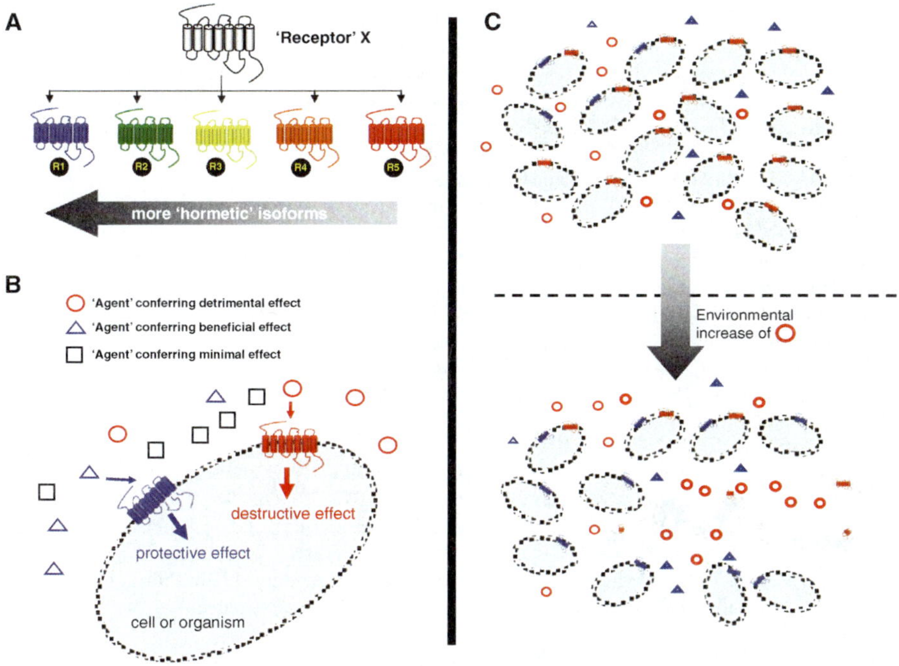

Fig. 2 GPCR isoform repertoire may provide a mechanism for hormetic effects and survival selection. **A.** The potential creation of multiple stable isoforms of a single rhodopsin-like receptor (R1–R5). These receptor forms are created by differential stable intramolecular interactions with distinct stoichiometries of accessory or scaffolding proteins. These isoforms also may exhibit a spectrum of strength of linkage to downstream signaling pathways that are more (R1) or less (R5) beneficial (or hormetic) to the survival of the cell/organism. **B.** A theoretical environment in which a cell or organism possesses the ability to sense various agents in its environment (*circle, triangle, square*). These agents possess the ability to activate either the beneficial or deleterious isoforms of a single receptor type (X). **C.** The potential effects on cell survival or organism propagation based on the sensory ability of the cell using receptor X. In an unstable environment in which extracellular agents may change concentration, survival of certain cells/organisms may depend both on receptor X expression and also on the specific isoforms of receptor X that exist. It is likely that retention of isoforms that confer beneficial effects and loss of isoforms that confer deleterious effects can occur simultaneously. In addition, the ability to produce and stabilize the most receptor X isoforms may confer future flexibility of the cell/organism to future changes in environment

References

Ali MS, Sayeski PP, Dirksen LB, et al. (1997) Dependence on the motif YIPP for the physical association of jak2 kinase with the intracellular carboxyl tail of the angiotensin II AT1 receptor. J Biol Chem 272: 23382–23388.

Angers S, Salahpour A, Bouvier M. (2002) Dimerization: an emerging concept for G protein–coupled receptor ontogeny and function. Annu Rev Pharmacol Toxicol 42: 409–435.

Arch JR (2004) Do low-affinity states of beta-adrenoceptors have roles in physiology and medicine? Br J Pharmacol. 143: 517–518.

Baker JG (2005) Site of action of beta-ligands at the human beta1-adrenoceptor. J Pharmacol Exp Ther 13: 1163–1171.

Berg KA, Maayani S, Goldfarb J, et al. (1998) Effector pathway–dependent relative efficacy at serotonin type 2a and 2c receptors: evidence for agonist-directed trafficking of receptor stimulus. Mol Pharmacol 54: 94–104.

Birdsall NJM, Hulme EC, Stockton JM (1983) Muscarinic receptor subclasses: allosteric interactions. Cold Spring Harb Symp Quant Biol 48: 53–56.

Birdsall NJM, Lazareno S (2005) Allosterism at muscarinic receptors: ligands and mechanisms. Mini-Rev Med Chem 5: 523–543.

Bockaert J, Marin P, Dumuis A, et al. (2003) The 'magic tail' of G protein–coupled receptors: an anchorage for functional protein networks. FEBS Lett 546: 65–72.

Brady AE, Limbird LE (2002) G protein–coupled receptor interacting proteins: emerging roles in localization and signal transduction. Cell Signal 14: 297–309.

Chadwick W, Magnus T, Martin B, et al. (2008) Targeting TNF-alpha receptors for neurotherapeutics. Trends Neurosci 31: 504–511.

Christopoulos A, Kenakin T (2002) G protein–coupled receptor allosterism and complexing. Pharmacol Rev 54: 323–374.

Christopoulos A, May LT, Avlani VA, et al. (2004) G-protein–coupled receptor allosterism: the promise and the problem(s). Biochem Soc Trans 32(Pt 5): 873–877.

Daaka Y, Luttrell LM, Lefkowitz RJ (1997) Switching of the coupling of the β2-adrenergic receptor to different G proteins by protein kinase A. Nature 390: 88–91.

DeLean A, Stadel JM, Lefkowitz RJ (1980) A ternary complex model explains the agonist-specific binding properties of the adenylate cyclase–coupled beta-adrenergic receptor. J Biol Chem 255: 7108–7117.

Franco R, Ferre S, Agnati L, et al. (2000) Evidence for adenosine/dopamine receptor interactions: indications for heteromerization. Neuropsychopharmacology 23: S50–S59.

Fredriksson R, Schiöth HB (2005) The repertoire of G-protein–coupled receptors in fully sequenced genomes. Mol Pharmacol 67: 1414–1425.

George SR, Fan T, Xie Z, et al. (2000) Oligomerization of μ- and δ-opioid receptors. Generation of novel functional properties. J Biol Chem 275: 26128–26135.

Granneman JG (2001) The putative β4-adrenergic receptor is a novel state of the β1-adrenergic receptor. Am J Physiol Endocrinol Metab 280: E199–E202.

Guo W, Shi L, Filizola M, et al. (2005) Crosstalk in G protein–coupled receptors: changes at the transmembrane homodimer interface determine activation. Proc Natl Acad Sci USA 102: 17495–17500.

Holloway AC, Qian H, Pipolo L, et al. (2002) Side-chain substitutions within angiotensin II reveal different requirements for signaling, internalization, and phosphorylation of type 1a angiotensin receptors. Mol Pharmacol 61: 768–777.

Jones KA, Borowsky B, Tamm JA, et al. (1998) GABA(B) receptors function as a heteromeric assembly of the subunits GABA(B)R1 and GABA(B)R2. Nature 396: 674–679.

Jordan BA, Devi LA (1999) G-protein–coupled receptor heterodimerization modulates receptor function. Nature 399: 697–700.

Keith DE, Murray SR, Zaki PA, et al. (1996) Morphine activates opioid receptors without causing their rapid internalization. J Biol Chem 271: 19021–19024.

Kenakin T (2002) Drug efficacy at G protein–coupled receptors. Ann Rev Pharmacol Toxicol 42: 349–379.

Kenakin T (2003) Ligand-selective receptor conformations revisited. The promise and the problem. Trends Pharmacol Sci 24: 346–354.

Kohout TA, Nicholas SL, Perry SJ, et al. (2004) Differential desensitization, receptor phosphorylation, β-arrestin recruitment and ERK1/2 activation by the two endogenous ligands for the CC chemokine receptor 7. J Biol Chem 279: 23214–23222.

Konkar AA (2000) Aryloxypropanolamine and catecholamine ligand interactions with the β1 adrenergic receptor: evidence for interaction with distinct conformations of β1-adrenergic receptors. J Pharmacol Exp Ther 294: 923–932.

Lazareno S (2004) Thiochrome enhances acetylcholine affinity at muscarinic M4 receptors: receptor subtype selectivity via cooperativity rather than affinity. Mol Pharmacol 65: 257–266.

Laporte SA, Oakley RH, Zhang J, et al. (1999) The beta2-adrenergic receptor/beta arrestin complex recruits the clathrin adaptor AP-2 during endocytosis. Proc Natl Acad Sci USA 96: 3712–3717.

Liang Y, Fotiadis D, Filipek S, et al. (2003) Organization of the G protein–coupled receptors rhodopsin and opsin in native membranes. J Biol Chem 278: 21655–21662.

Lowe MD (2002) Comparison of the affinity of β-blockers for the two states of the β1-adrenoceptor in ferret ventricular myocardium. Br J Pharmacol 135: 451–461.

Luttrell LM, Ferguson SSG, Daaka Y et al. (1999) □-Arrestin–dependent formation of β2 adrenergic receptor/src protein kinase complexes. Science 283: 655–661.

Marlo JE, Niswender CM, Days EL, et al. 2008 Discovery and characterization of novel allosteric potentiators of M1 muscarinic receptors reveals multiple modes of activity. Mol Pharmacol Dec 1 [Epub ahead of print].

Maudsley S, Gent JP, Findlay JB, et al. (1998) The relationship between the agonist-induced activation and desensitization of the human tachykinin NK2 receptor expressed in *Xenopus* oocytes. Br J Pharmacol 124: 675–684.

Maudsley S, Pierce KL, Zamah AM, et al. (2000) The beta(2)-adrenergic receptor mediates extracellular signal-regulated kinase activation via assembly of a multi-receptor complex with the epidermal growth factor receptor. J Biol Chem 275: 9572–9580.

Maudsley S, Zamah AM, Rahman N, Blitzer JT, Luttrell LM, Lefkowitz RJ, Hall RA (2000a) Platelet-derived growth factor receptor association with Na(+)/H(+) exchanger regulatory factor potentiates receptor activity. Mol Cell Biol 20: 8352–8363.

Maudsley S, Davidson L, Pawson AJ, et al. (2004) Gonadotropin-releasing hormone (GnRH) antagonists promote proapoptotic signaling in peripheral reproductive tumor cells by activating a galphai-coupling state of the type I GnRH receptor. Cancer Res 64: 7533–7544.

Maudsley S, Martin B, Luttrell LM (2005) The origins of diversity and specicity in G protein coupled receptor signaling. J Pharmacol Exp Ther 314: 485–494.

Maudsley S, Martin B, Luttrell LM (2007b) G protein–coupled receptor signaling complexity in neuronal tissue: implications for novel therapeutics. Curr Alzheimer Res 4: 3–19.

Maudsley S, Naor Z, Bonfil D, et al. (2007a) Proline-rich tyrosine kinase 2 mediates gonadotropin-releasing hormone signaling to a specific extracellularly regulated kinase-sensitive transcriptional locus in the luteinizing hormone beta-subunit gene. Mol Endocrinol 21: 1216–1233.

May LT, Leach K, Sexton PA, et al. (2007) Allosteric modulation of g protein–coupled receptors ann. Rev Pharmacol Toxicol 47: 1–51.

Moolenaar P (2003) The 'state' of β-adrenoceptors. Br J Pharmacol 140: 1–2.

Pak MD, Fishman PH (1996) Anomalous behavior of CGP 12177a on β1-adrenergic receptors. J Recept Signal Transduct Res 16: 1–23.

Peroutka SJ, Snyder SH (1980) Regulation of serotonin2 (5-HT2) receptors labeled with [^{3}H]spiroperidol by chronic treatment with the antidepressant amitriptyline. J Pharmacol Exp Ther 215: 582–587.

Pommier B, Da Nascimento S, Dumont S, et al. (1999) The cholecystokinin B receptor is coupled to two effector pathways through pertussis toxin–sensitive and –insensitive G proteins. J Neurochem 73: 281–288.

Rimoldi V, Reversi A, Taverna E, et al. (2003) Oxytocin receptor elicits different EGFR/MAPK activation patterns depending on its localization in caveolin-1 enriched domains. Oncogene 22: 6054–6060.

Roettger BF, Ghanekar D, Rao R, Toledo C, Yingling J, Pinon D,, Miller LJ (1997) Antagonist-stimulated internalization of the G protein–coupled cholecystokinin receptor. Mol Pharmacol 51: 357–362.

Sagan S, Karoyan P, Chassaing G, et al. (1996) Tachykinin peptides affect differently the second messenger pathways after binding to CHO-expressed human NK-1 receptors. J Pharmacol Exp Ther 276: 1039–1048.

Samama P, Cotecchia S, Costa T, et al. (1993) A mutation-induced activated state of the beta 2-adrenergic receptor. Extending the ternary complex model. J Biol Chem 268: 4625–4636.

Sexton PM, Albiston A, Morfis M, et al. (2001) Receptor activity modifying proteins. Cell Signal 13: 73–83.
Sneddon WB, Syme CA, Bisello A, et al. (2003) Activation-independent parathyroid hormone receptor internalization is regulated by NHERF1 (EBP50). J Biol Chem 278: 43787–43796.
Soudijn W (2004) Allosteric modulation of G protein–coupled receptors: perspective and recent developments. Drug Discov Today 9: 752–758.
Stout BD, Clarke WP (2002) Berg KA rapid desensitization of the serotonin(2c) receptor system: effector pathway and agonist dependence. J Pharmacol Exp Ther 302: 957–962.
Takeda S, Kadowaki S, Haga T, et al. (2002) Identification of G protein–coupled receptor genes from the human genome sequence. FEBS Lett 520: 97–101.
Tang Y, Hu LA, Miller WE, et al. (1999) Identification of the endophilins (SH3p4/p8/p13) as novel binding partners for the beta1-adrenergic receptor. Proc Natl Acad Sci USA 96: 12559–12564.
Urban JD, Clarke WP, von Zastrow M, et al. (2007) Functional selectivity and classical concepts of quantitative pharmacology. J Pharmacol Exp Ther 320: 1–13.
van Hooft JA, Vijverberg HP (1996) Selection of distinct conformational states of the 5-HT3 receptor by full and partial agonists. Br J Pharmacol 117: 839–846.
Waelbroeck M (1994) Identification of drugs competing with d-tubocurarine for an allosteric site on cardiac muscarinic receptors. Mol Pharmacol 46: 685–692.
Wang Q, Zhao J, Brady AE et al (2004) Spinophilin blocks arrestin actions in vitro and in vivo at G protein–coupled receptors. Science 304: 1940–1944.
Watson C (2005) The CCR5 receptor–based mechanism of action of 873140, a potent allosteric non competitive HIV entry inhibitor. Mol Pharmacol 67: 1268–1282.
Yu Y, Zhang L, Yin X, et al. (1996) μ-Opioid receptor phosphorylation, desensitization and ligand efficacy. J Biol Chem 272: 28869–28874.
Zhang SJ, Cheng H, Zhou YY, et al. (2000) Inhibition of spontaneous beta 2-adrenergic activation rescues beta 1-adrenergic contractile response in cardiomyocytes overexpressing beta 2-adrenoceptor. J Biol Chem 275: 21773–21779.

Exercise-Induced Hormesis

Alexis M. Stranahan and Mark P. Mattson

Abstract The consequences of physical activity on the brain can readily be integrated into a hormetic framework. Whereas low- to moderate-intensity exercise exerts positive effects on the body, excessive exercise can be detrimental for somatic health. Here we review the evidence linking physical activity with cellular and functional modifications in different organ systems, with a focus on the dose-response characteristics of this relationship. Voluntary running and short-term treadmill running within the range of intensities normally experienced during voluntary running both enhance metabolism and preserve function across multiple organ systems. In contrast, running to exhaustion has a negative impact on global functioning. Overall, the effects of exercise clearly depend on the amount and intensity of activity. These effects conform to the biological principle of hormesis.

Keywords Running · Exercise · Wheel · Treadmill · Hippocampus

Introduction

Physical activity induces dose-dependent physiological adaptations in virtually every organ system of the body. The "overload principle" governs progressive adaptation to increasing physiological demands over time: Based on this principle, gradual increases in the physiological demand for energy with exercise training result in adaptive mechanisms that increase metabolic efficiency (Rhea and Alderman, 2004). In contrast with the adaptive mechanisms induced by progressive overload, an acute bout of exercise to exhaustion frequently activates mechanisms that oppose those of gradual increases in training intensity.

A.M. Stranahan (✉)
Department of Psychological and Brain Sciences, Johns Hopkins University, Baltimore, MD 21218, USA
e-mail: alexis.stranahan@jhu.edu

M.P. Mattson, E.J. Calabrese, *Hormesis*, DOI 10.1007/978-1-60761-495-1_6,

The dose-dependent nature of physiological adaptation to exercise has been extensively studied in animal models. In rodent models, "moderate" exercise is generally induced using either (1) treadmill training at low speeds using minimal incline or (2) voluntary wheel running. Although these two exercise modalities do differ, it is generally accepted that both moderate treadmill training and voluntary running induce adaptive physiological responses in rodents. Exhaustive exercise is most often induced using treadmill running at high speeds, sometimes using additional incline; some studies have also modeled exhaustive exercise using food restriction–induced hyperactivity in the running wheel. Both treadmill running to exhaustion and food restriction–induced wheel running have been shown to induce maladaptive responses in rodents.

In this chapter, we review evidence for a hormetic dose-response relationship between exercise and physiological adaptation in different organ systems. To afford greater insight into the biological mechanisms induced by different training modalities, the data that we review are derived from studies in animal models. The global principles derived from these studies suggest that most organ systems conform to the biological principle of hormesis with respect to the adaptive and maladaptive response to different exercise training regimens.

Effects of Exercise on the Musculoskeletal System

Dose-Response Characteristics of Exercise Effects on Muscle

In rodent models, wheel running and treadmill running are accompanied by structural alterations in skeletal muscle. The dose-dependent effects of exercise are well recognized in this somatic compartment; high-frequency, high-intensity training is associated with muscle breakdown, whereas moderate-frequency, moderate-intensity training is associated with increased vascularization of the muscle (Baar, 2006). The positive consequences of endurance training are associated with improved cellular calcium metabolism (Chin, 2005) and increased antioxidant capacity (Caillaud et al., 1999). Finally, resident stem cells in muscle tissue respond to moderate exercise with increased proliferation, contributing to muscular hypertrophy (Choi et al., 2005).

Repeated training to exhaustion with insufficient rest leads to myocyte apoptosis and, over time, activates catabolic mechanisms leading to loss of muscle tissue (Steinacker et al., 2004). Moderate exercise, such as that performed by rodents running voluntarily in a wheel, exerts positive effects on skeletal muscle (Choi et al., 2005), whereas repeated, high-intensity exercise, such as that experienced by rodents subjected to repeated bouts of high-intensity treadmill running, induces muscular breakdown (Matsunaga et al., 2007). In the musculoskeletal system more than in any other somatic compartment, it is clear that the effects of exercise conform to the principle of hormesis.

Duration-Dependent Effects of Exercise on Bone

The relationship between exercise and the musculoskeletal system is also evident at the skeletal level. Voluntary wheel running increased bone strength (Warren et al., 2007), whereas food restriction–induced hyperactivity in the wheel was associated with reduced bone mineral density (Dimarco et al., 2007). Thus, the dose-response relationship between exercise and cellular integrity appears to be a general feature of cells in different tissues of the musculoskeletal system.

Effects of Exercise on the Digestive System

Dose-Dependent Effects of Exercise on the Stomach

Ghrelin is an orexigenic peptide produced in the stomach (Inui et al., 2004). Concentrations of ghrelin in serum were reduced following exercise training (Ebal et al., 2007). Because ghrelin is negatively regulated by growth hormone, a reduction in serum ghrelin concentrations may be indicative of improved growth hormone negative feedback. In this manner, it is possible that moderate exercise improves growth hormone negative feedback in the stomach. In contrast with the positive effects of moderate exercise, strenuous exercise has been shown to promote the formation of gastric ulcers (Pare and Houser, 1973). This suggests that the dose-dependent effects of exercise on the stomach conform to the principle of hormesis.

Hormetic Effects of Exercise in the Large and Small Intestine

The hormetic effects of exercise are also apparent in the intestinal compartment. Moderate exercise suppresses the transcription of genes that have been linked with colon cancer (Buehlmeyer et al., 2008). Although moderate exercise reduces oxidative stress in the rodent intestine (de Lira et al., 2008), high-intensity exercise disrupts the integrity of intestinal mitochondria (Rosa et al., 2008). This suggests that the intestinal response to physical activity also has dose- and duration-dependent components.

Dose-Response Characteristics of Exercise Effects on the Liver

While moderate exercise reduces oxidative stress levels, strenuous exercise is associated with increased oxidative stress. These effects are especially apparent in the liver. Changes in oxidative capacity with moderate exercise were associated with an increase in mitochondrial complex IV activity (Boveris and Navarro, 2008). In a detailed study using moderate, intermediate, or strenuous treadmill training,

Ogonovszky et al. (2005) observed increased oxidative damage in the liver following strenuous training. Similarly, treadmill running to exhaustion was associated with increased concentrations of malondialdehyde, an oxidative stress marker, in the liver (Liu et al., 2000). These studies point to hormetic effects of exercise, particularly with respect to hepatic mitochondrial function.

Duration-Dependent Effects of Exercise on the Pancreas

Moderate exercise increases pancreatic volume and β-cell mass (Shima et al., 1997) and reduces circulating insulin levels, indicative of enhanced insulin sensitivity (Ebal et al., 2007). Pancreatic β-cells also express neuronal nitric oxide synthase (nNOS), and nNOS expression in this cell type is increased by moderate exercise (Ueda et al., 2003). Moderate exercise also slows disease progression in animal models of insulin-resistant diabetes (Stranahan et al., 2009; Shima et al., 1997).

In contrast with the positive consequences of moderate exercise, treadmill running to exhaustion exerts deleterious effects. Treadmill running to exhaustion induces hypoglycemia, indicative of an altered balance between glucose and insulin levels (Zendzian-Piotrowska and Górski, 1993). Because exogenous glucose extends run time to exhaustion (Arogyasami et al., 1992), exercise-induced hypoglycemia may contribute to overall fatigue. In this regard, strenuous exercise perturbs pancreatic homeostasis, while moderate exercise improves insulin and glucose homeostasis, in part through direct effects on the pancreas.

Effects of Exercise on the Reproductive System

Dose-Response Characteristics of Exercise Effects on Ovarian Function

The female reproductive cycle is particularly susceptible to disruption by strenuous exercise. In female rats, voluntary running or treadmill training both cause irregularities in the estrus cycle in a subset of animals (Chatterton et al., 1995; Caston et al., 1995). However, this may be offset by increased production of sex steroids in other tissues, such as the muscle (Aizawa et al., 2008). Because the female reproductive system is dependent on the metabolic status of the animal, the hormetic effects of exercise should be considered with respect to overall calorie sensing rather than activity levels (Martin et al., 2007). Of interest, when faced with reduced dietary energy availability, female rats increased their levels of spontaneous activity, ceased estrous cycling, underwent endocrine masculinization, exhibited a heightened stress response, and improved their learning and memory (Martin et al., 2007). These findings suggest a conserved evolutionary basis for the increased spontaneous exercise in women with anorexia nervosa.

Duration-Dependent Effects of Exercise on the Testis

The male reproductive system fits the theory of hormesis better than the female reproductive system. Moderate exercise training reduces oxidative stress levels in the testis (Husain and Somani, 1998), whereas acute running to exhaustion increases testicular oxidative stress levels (Aksoy et al., 2006). Exercise also increases levels of the antioxidant glutathione in the testis of aged rats (Somani and Husain, 1996). In a recent study it was reported that lifelong exercise (voluntary wheel running) protects the testes of mice against the adverse effects of aging, as indicated by the preservation of cellular composition, including a full complement of cells at different stages of the spermatogenic process, and by decreased amounts of oxidative damage to spermatogenic and Leydig cells (Chigurupati et al., 2008).

Wheel running was associated with elevated levels of testosterone and follicle-stimulating hormone in the serum of male rodents (Pieper et al., 1995). This observation may not hold true for treadmill training, which was associated with increased in vitro production of testosterone by Leydig cells, with no change in circulating serum concentrations (Härkönen et al., 1990). In contrast with the effects of voluntary running or moderate treadmill training, treadmill running to exhaustion reduces testosterone levels (Guezennec et al., 1982). With respect to oxidative stress levels and testosterone production, it is clear that the effects of exercise on male reproductive function conform to the principle of hormesis.

Effects of Exercise on the Cardiorespiratory System

Duration-Dependent Effects of Exercise on the Heart

Running promotes cardiac hypertrophy and enhances the function of the heart and circulatory system (Watson et al., 2007). Moderate voluntary running increases the number and volume of mitochondria in heart cells (Eisele et al., 2008). Lower-intensity treadmill training also enhances cardiac calcium metabolism and increases contractile strength in cardiac muscle (Kemi et al., 2007). Overall, it is apparent that moderate exercise is cardioprotective.

The dose-response curve for the effects of exercise in the heart is shifted to the right of that for most other organs. There is very little evidence to suggest that running to exhaustion exerts any negative effects on the heart. One compensatory mechanism that may account for the lack of any negative effects of exercise to exhaustion involves the running-induced induction of heat-shock proteins in the heart and other organs (Campisi et al., 2003). Because heat-shock proteins buffer the deleterious consequences of cellular stress, their induction may account for the resistance of the heart to the negative consequences of running to exhaustion.

Alternatively, because most studies of running to exhaustion use behavioral criteria to determine "exhaustion," it is possible that cardiac cells are simply subjected to less stress than skeletal muscles or are more resistant to the effects of strenuous exercise.

Dose-Dependent Effects of Exercise on the Lungs

Running is widely accepted as enhancing maximal oxygen uptake, which is an indicator of improved function of the respiratory system (Swallow et al., 1998). Moderate exercise also enhances antioxidant capacity in the lungs (Duncan et al., 1997) and reduces the formation of lung tumors (Murphy et al., 2004). Although it is clear that moderate running enhances respiratory function, there is very little indication that running to exhaustion exerts detrimental effects on the lungs.

Effects of Exercise on the Immune System

Dose-Response Characteristics of Exercise Effects on the Thymus

The thymus plays a central role in adaptive immunity. T-cells produced in the thymus mediate the mnemonic component of the immune response, and the mobilization of T-cells is enhanced by moderate exercise (Krüger et al., 2008). Before differentiating into mature T-cells, bone marrow stem cells go through a transient stage of differentiation as thymocytes, and moderate exercise also increases the number of thymocytes in rats (Ferry et al., 1992). Through these and other mechanisms, moderate exercise training enhances adaptive immunity.

The consequences of exhaustive exercise oppose those of moderate training. Treadmill running to exhaustion induces programmed cell death among thymocytes (Concordet and Ferry, 1993; Quadrilatero and Hoffman-Goetz, 2005). Exhaustive exercise also impairs the T-cell response to infection (Kapasi et al., 2005; Kohut et al., 2001). In this regard, exhaustive exercise impairs, while moderate exercise enhances, thymic function.

Hormetic Effects of Exercise on the Spleen

The spleen is also responsive to the hormetic effects of exercise. While moderate exercise increases antiapoptotic protein expression in splenic lymphocytes (Hoffman-Goetz and Spagnuolo, 2007; Avula et al., 2001), intense exercise reduces the production of splenic T-lymphocytes (Hoffman-Goetz et al., 1986). Moderate exercise training also attenuates the deleterious effects of psychological stress on splenic function (Dishman et al., 2000). In addition to blunting the effects of psychological stress, moderate exercise training also restores glucose uptake in splenic lymphocytes from diabetic rodents (Moriguchi et al., 1998). Based on these data, it is apparent that exercise exerts dose-dependent effects on splenic contributions to immunity, and moderate exercise protects the immune system against a variety of psychological and physiological insults.

Exercise Effects on Circulating Cytokines Conform to the Theory of Hormesis

Physical activity also influences a variety of serum parameters involved in immune regulation. Moderate voluntary exercise increased levels of natural immunoglobulin M (IgM) in serum (Elphick et al., 2003) and enhanced the expression of serum interleukin-10 following an immune challenge in mice (Kohut et al., 2004; Nickerson et al., 2005). In contrast, exhaustive treadmill exercise suppresses the secondary antibody response to infection (Kapasi et al., 2005). In this regard, it is apparent that moderate exercise enhances, while strenuous exercise impairs, serum markers of active and innate immunity in rodents. Moderate exercise can increase macrophage activity, and catecholamines may mediate this enhancement of innate immunity (Ortega et al., 2007). Macrophage resistance to herpes simplex virus 1 (HSV-1) was increased with both exercise and oat beta-glucan, whereas natural killer (NK) cell cytotoxicity was only increased with exercise. Exercise was also associated with 45% and 38% decreases in morbidity and mortality, respectively. Mortality was also decreased with oat beta-glucan, but this effect did not reach statistical significance. No additive effects of exercise and oat beta-glucan were found. These data confirm a positive effect of both moderate exercise and oat beta-glucan on immune function, but only moderate exercise was associated with a significant reduction in the risk of upper respiratory tract infection in this model. Similarly, moderate exercise enhances innate cellular immune responses in the lungs, thereby increasing resistance to respiratory infection (Davis et al., 2004). On the other hand, after competing in a marathon, runners typically exhibit reduced NK cell, neutrophil, and macrophage activity for up to 3 days following the event, which may increase their vulnerability to infection (Nieman, 2007).

Effects of Exercise on the Brain

Duration-Dependent Effects of Exercise on Adult Neurogenesis

Voluntary wheel running promotes the formation of newly generated neurons and astrocytes in the adult dentate gyrus of the hippocampus, a brain region that plays a critical role in learning and memory. This effect has been reported following as little as 9 days of training (van Praag et al., 1999). There does not appear to be a progressive increase in the number of newly generated cells with longer periods of activity; in contrast, the effect resembles a switch, with production of new cells occurring at a higher level that is consistent over longer periods. The underlying mechanism for this switch may involve faster cycling among an existing pool of progenitor cells, or an additional pool of quiescent progenitors may be recruited into the proliferative population.

The effects of running encompass both cell proliferation and cell survival. It has yet to be determined whether similar mechanisms are involved in promoting the generation and survival of new neurons and astrocytes in the brains of runners. Moreover, no studies have systematically addressed whether effects on survival are switch-like, similar to effects on proliferation, or whether there may be a cumulative enhancement of cell survival over time. The observation that the volume of the dentate gyrus increases over time (Stranahan et al., 2006) would seem to favor the second, cumulative hypothesis.

Running also influences the intrinsic capacity for structural plasticity among newly generated neurons. Specifically, new neurons in the brains of runners, visualized with retroviral labeling, show greater plasticity among dendritic spines relative to new neurons in the brains of sedentary mice (Zhao et al., 2006). Despite this enhancement of structural plasticity in the brains of exercising mice, there was no effect of running on the intrinsic electrophysiological properties of newly generated neurons (Jakubs et al., 2006). However, an increase in the number of new neurons, with their lower firing thresholds and altered calcium metabolism, may well be sufficient to influence the circuit properties of the dentate gyrus as a whole.

Rats normally run in bouts of 5 to 7 seconds each (Stranahan et al., 2006). These frequent, high-intensity intervals of running typically amount to between 2 and 7 km/day for Sprague-Dawley rats. Studies in which increased neurogenesis was reported following treadmill training typically used lower-intensity exercise, between 0.7 and 1.0 km/day (Trejo et al., 2001; Llorens-Martin et al., 2006). The observation that 8 days of forced running impairs adult neurogenesis fits well within the theory of hormesis (Roman et al., 2005). This comparison suggests that lower-intensity involuntary exercise or moderate-intensity voluntary exercise exerts beneficial effects on the hippocampus, while higher-intensity forced running over longer periods has detrimental consequences for hippocampal plasticity.

Dose-Response Characteristics of Exercise Effects on Dendritic Spines

New neurons are not unique, in that they respond to wheel running with enhanced plasticity among spines. The running-induced enhancement of dendritic spine plasticity is a general characteristic for both newly generated and mature dentate gyrus granule cells.

Several studies have now shown that running increases dendritic spine density among dentate gyrus granule cells. Although far fewer studies have looked at the CA1 field, there also has been some indication of enhanced spinogenesis in this hippocampal subfield. Specifically, CA1 pyramidal neurons respond to wheel running with increased dendritic spine density and also exhibit a bias in favor of longer, thinner spines (Stranahan et al., 2007). These observations suggest that the enhancement of dendritic plasticity by wheel running is a general feature of neurons in the hippocampus.

Few studies have addressed the possibility of duration- or intensity-dependent effects of wheel running on spinogenesis. Increased dendritic spine density has been reported in the dentate gyrus following 2 weeks (Eadie et al., 2005; Redila and Christie, 2006) or 8 weeks (Stranahan et al., 2007) of running; again, this rapid onset of spinogenesis is suggestive of a switch-like mechanism rather than a cumulative enhancement of spine density over time. There is some indication that excessive running may be detrimental for neuronal dendrites; specifically, food restriction–induced hyperactivity in the running wheel is accompanied by dendritic retraction and loss of spines in the hippocampus (Lambert et al., 1998). However, additional studies will be necessary to determine whether these effects are due to excessive running or inadequate nutrition.

Duration-Dependent Effects on Angiogenesis

In addition to increasing the production of new neurons and astrocytes, physical activity also promotes angiogenesis across multiple brain regions. The earliest report of exercise-induced angiogenesis came from the work of Greenough and colleagues (Isaacs et al., 1992), who observed increased capillary density in the cerebellum following physical activity. Later studies also reported increased vascular capacity in the motor cortex and striatum (McCloskey et al., 2001), as well as in the hippocampus (Lopez-Lopez et al., 2004; but see McCloskey et al., 2001). With regard to hematopoietic stem cells, wheel running also promotes increased proliferation in this cell population, in addition to modulating stem cell proliferation in the brain (Laufs et al., 2004).

The dose-response characteristics of the relationship between exercise and central neovascularization has not been addressed experimentally. It is also difficult to generalize across studies because the above reports employed wheel running (McCloskey et al., 2001; Isaacs et al., 1992) or treadmill training (Lopez-Lopez et al., 2004; Isaacs et al., 1992) as a means of increasing physical activity. The possibility of a hormetic response to exercise in the angiogenic compartment will be a fruitful avenue for future study.

Duration-Dependent Effects of Running on Neurotrophic Factor Expression

Wheel running increases the expression of a variety of neurotrophic factors, including brain-derived neurotrophic factor (BDNF), insulin-like growth factor 1 (IGF-1), vascular endothelial growth factor (VEGF), and neurotrophin-3 (NT-3) (for review, see Cotman et al., 2007). Among these growth factors, the best-characterized dose-response relationship is the one between exercise and BDNF levels in the hippocampus. Adlard and colleagues (2004) outlined the time course for the enhancement of BDNF levels following voluntary wheel running, and their research suggested that BDNF protein levels are elevated following 4 weeks of running. However, no studies have determined whether excessive exercise might actually reduce levels of BDNF or other neurotrophic factors.

Conclusion

Based on these studies, several conclusions can be drawn regarding the dose-response characteristics of the effects of exercise on different organ systems. First, the heart and lungs can tolerate higher exercise intensities than other organ systems (Fig. 1). Second, female reproductive function exhibits a hormetic relationship with overall calorie balance rather than levels of physical activity. Third, the musculoskeletal system is the best-characterized example demonstrating trophic effects of moderate exercise and detrimental effects of severe exhaustive exercise, followed by the central nervous system, immune system, and male reproductive system. Finally, exercise effects on digestive function are less well understood, although at first glance they may appear to conform to the principle of hormesis.

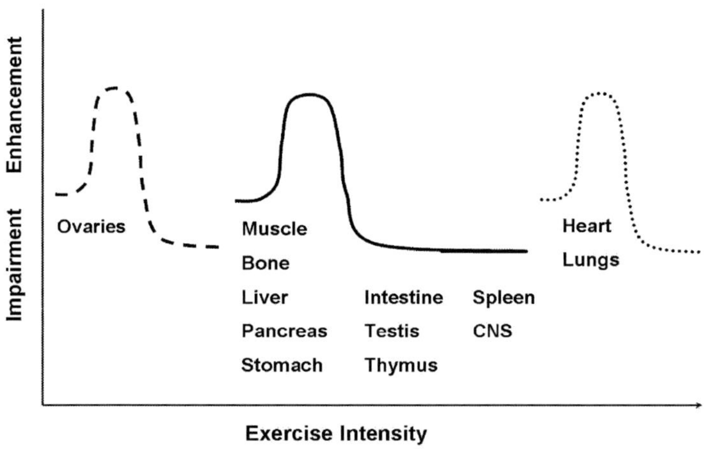

Fig. 1 Variability in the threshold for the hormetic effects of exercise in different organ systems. CNS, central nervous system

Acknowledgments This work was supported by the Intramural Research Program of the National Institute on Aging, National Institutes of Health.

References

Adlard PA, Perreau VM, Engesser-Cesar C, Cotman CW (2004) The timecourse of induction of brain-derived neurotrophic factor mRNA and protein in the rat hippocampus following voluntary exercise. Neurosci Lett 363: 43–48.

Aizawa K, Iemitsu M, Otsuki T, Maeda S, Miyauchi T, Mesaki N (2008) Sex differences in steroidogenesis in skeletal muscle following a single bout of exercise in rats. J Appl Physiol 104: 67–74.

Aksoy Y, Yapanoğlu T, Aksoy H, Demircan B, Oztaşan N, Canakçi E, Malkoçl (2006) Effects of endurance training on antioxidant defense mechanisms and lipid peroxidation in testis of rats. Arch Androl 52: 319–323.

Arogyasami J, Sellers TL, Wilson GI, Jones JP, Duan C, Winder WW (1992) Insulin-induced hypoglycemia in fed and fasted exercising rats. J Appl Physiol 72: 1991–1998.

Avula CP, Muthukumar AR, Zaman K, McCarter R, Fernandes G (2001) Inhibitory effects of voluntary wheel exercise on apoptosis in splenic lymphocyte subsets of C57BL/6 mice. J Appl Physiol 91: 2546–2552.

Baar K (2006) Training for endurance and strength: lessons from cell signaling. Med Sci Sports Exerc 38: 1939–1944.

Boveris A, Navarro A (2008) Systemic and mitochondrial adaptive responses to moderate exercise in rodents. Free Radic Biol Med 44: 224–229.

Buehlmeyer K, Doering F, Daniel H, Kindermann B, Schulz T, Michna H (2008) Alteration of gene expression in rat colon mucosa after exercise. Ann Anat 190: 71–80.

Caillaud C, Py G, Eydoux N, Legros P, Prefaut C, Mercier J (1999) Antioxidants and mitochondrial respiration in lung, diaphragm, and locomotor muscles: effect of exercise. Free Radic Biol Med 26: 1292–1299.

Campisi J, Leem TH, Greenwood BN, Hansen MK, Moraska A, Higgins K, Smith TP, Fleshner M (2003) Habitual physical activity facilitates stress-induced HSP72 induction in brain, peripheral, and immune tissues. Am J Physiol Regul Integr Comp Physiol 284: R520–R530.

Caston AL, Farrell PA, Deaver DR (1995) Exercise training-induced changes in anterior pituitary gonadotrope of the female rat. J Appl Physiol 79: 194–201.

Chatterton RT Jr, Hrycyk L, Hickson RC(1995) Effect of endurance exercise on ovulation in the rat. Med Sci Sports Exerc 27: 1509–1515.

Chigurupati S, Son TG, Hyun DH, Lathia JD, Mughal MR, Savell J, Li SC, Nagaraju GP, Chan SL, Arumugam TV, Mattson MP (2008) Lifelong running reduces oxidative stress and degenerative changes in the testes of mice. J Endocrinol 199: 333–341.

Chin ER (2005). Role of Ca2+/calmodulin-dependent kinases in skeletal muscle plasticity. J Appl Physiol 99(2): 414–423.

Choi S, Liu X, Li P, Akimoto T, Lee SY, Zhang M, Yan Z (2005) Transcriptional profiling in mouse skeletal muscle following a single bout of voluntary running: evidence of increased cell proliferation. J Appl Physiol 99: 2406–2415.

Concordet JP, Ferry A (1993) Physiological programmed cell death in thymocytes is induced by physical stress (exercise). Am J Physiol 265: C626–629.

Cotman CW, Berchtold NC, Christie LA (2007) Exercise builds brain health: key roles of growth factor cascades and inflammation. Trends Neurosci 30: 464–472.

Davis JM, Murphy EA, Brown AS, Carmichael MD, Ghaffar A, Mayer EP (2004) Effects of moderate exercise and oat beta-glucan on innate immune function and susceptibility to respiratory infection. Am J Physiol Regul Integr Comp Physiol 286: R366–372.

de Lira CA, Vancini RL, Ihara SS, da Silva AC, Aboulafia J, Nouailhetas VL (2008) Aerobic exercise affects C57BL/6 murine intestinal contractile function. Eur J Appl Physiol 103: 215–223.

Dimarco NM, Dart L, Sanborn CB (2007) Modified activity-stress paradigm in an animal model of the female athlete triad. J Appl Physiol 103: 1469–1478.

Dishman RK, Hong S, Soares J, Edwards GL, Bunnell BN, Jaso-Friedmann L, Evans DL (2000) Activity-wheel running blunts suppression of splenic natural killer cell cytotoxicity after sympathectomy and footshock. Physiol Behav 71: 297–304.

Duncan K, Harris S, Ardies CM (1997) Running exercise may reduce risk for lung and liver cancer by inducing activity of antioxidant and phase II enzymes. Cancer Lett 116: 151–158.

Eadie BD, Redila VA, Christie BR (2005) Voluntary exercise alters the cytoarchitecture of the adult dentate gyrus by increasing cellular proliferation, dendritic complexity, and spine density. J Comp Neurol 486: 39–47.

Ebal E, Cavalie H, Michaux O, Lac G (2007) Effect of a moderate exercise on the regulatory hormones of food intake in rats. Appetite 49: 521–524.

Eisele JC, Schaefer IM, Randel Nyengaard J, Post H, Liebetanz D, Brüel A, Mühlfeld C (2008) Effect of voluntary exercise on number and volume of cardiomyocytes and their mitochondria in the mouse left ventricle. Basic Res Cardiol 103: 12–21.

Elphick GF, Wieseler-Frank J, Greenwood BN, Campisi J, Fleshner M (2003) B-1 cell (CD5+/CD11b+) numbers and nIgM levels are elevated in physically active vs. sedentary rats. J Appl Physiol 95: 199–206.

Ferry A, Rieu P, Laziri F, el Habazi A, Le Page C, Rieu M. (1992) Effect of moderate exercise on rat T-cells. Eur J Appl Physiol Occup Physiol 65: 464–468.

Guezennec CY, Ferre P, Serrurier B, Merino D, Pesquies PC (1982) Effects of prolonged physical exercise and fasting upon plasma testosterone level in rats. Eur J Appl Physiol Occup Physiol 49: 159–168.

Härkönen M, Näveri H, Kuoppasalmi K, Huhtaniemi I (1990) Pituitary and gonadal function during physical exercise in the male rat. J Steroid Biochem 35: 127–132.

Hoffman-Goetz L, Keir R, Thorne R, Houston ME, Young C (1986) Chronic exercise stress in mice depresses splenic T lymphocyte mitogenesis in vitro. Clin Exp Immunol 66: 551–557.

Hoffman-Goetz L, Spagnuolo PA (2007) Freewheel exercise training modifies pro- and anti-apoptotic protein expression in mouse splenic lymphocytes. Int J Sports Med 28: 787–791.

Husain K, Somani SM (1998) Interaction of exercise training and chronic ethanol ingestion on testicular antioxidant system in rat. J Appl Toxicol 18: 421–429.

Inui A, Asakawa A, Bowers CY, et al (2004). Ghrelin, appetite, and gastric motility: the emerging role of the stomach as an endocrine organ. FASEB J 18(3): 439–456.

Isaacs KR, Anderson BJ, Alcantara AA, Black JE, Greenough WT (1992) Exercise and the brain: angiogenesis in the adult rat cerebellum after vigorous physical activity and motor skill learning. J Cereb Blood Flow Metab 12: 110–119.

Jakubs K, Nanobashvili A, Bonde S, Ekdahl CT, Kokaia Z, Kokaia M, Lindvall O. (2006) Environment matters: synaptic properties of neurons born in the epileptic adult brain develop to reduce excitability. Neuron 52: 1047–1059.

Kapasi ZF, McRae ML, Ahmed R (2005) Suppression of viral specific primary T-cell response following intense physical exercise in young but not old mice. J Appl Physiol 98: 663–671.

Kemi OJ, Ellingsen O, Ceci M, Grimaldi S, Smith GL, Condorelli G, Wisløff U (2007) Aerobic interval training enhances cardiomyocyte contractility and Ca2+ cycling by phosphorylation of CaMKII and Thr-17 of phospholamban. J Mol Cell Cardiol 43: 354–361.

Kohut ML, Boehm GW, Moynihan JA (2001) Prolonged exercise suppresses antigen-specific cytokine response to upper respiratory infection. J Appl Physiol 90: 678–684.

Kohut ML, Thompson JR, Lee W, Cunnick JE (2004) Exercise training-induced adaptations of immune response are mediated by beta-adrenergic receptors in aged but not young mice. J Appl Physiol 96: 1312–1322.

Krüger K, Lechtermann A, Fobker M, Völker K, Mooren FC (2008) Exercise-induced redistribution of T lymphocytes is regulated by adrenergic mechanisms. Brain Behav Immun 22: 324–338.

Lambert KG, Buckelew SK, Staffiso-Sandoz G, Gaffga S, Carpenter W, Fisher J, Kinsley CH (1998) Activity-stress induces atrophy of apical dendrites of hippocampal pyramidal neurons in male rats. Physiol Behav 65: 43–49.

Laufs U, Werner N, Link A, Endres M, Wassmann S, Jürgens K, Miche E, Böhm M, Nickenig G (2004) Physical training increases endothelial progenitor cells, inhibits neointima formation, and enhances angiogenesis. Circulation 109: 220–226.
Liu J, Yeo HC, Overvik-Douki E, Hagen T, Doniger SJ, Chyu DW, Brooks GA, Ames BN (2000) Chronically and acutely exercised rats: biomarkers of oxidative stress and endogenous antioxidants. J Appl Physiol 89: 21–28.
Llorens-Martin M, Torres-Aleman I, Trejo JL (2006) Pronounced individual variation in the response to the stimulatory action of exercise on immature hippocampal neurons. Hippocampus 16: 480–490.
Lopez-Lopez C, LeRoith D, Torres-Aleman I (2004) Insulin-like growth factor I is required for vessel remodeling in the adult brain. Proc Natl Acad Sci USA 101: 9833–9888.
Martin B, Pearson M, Kebejian L, Golden E, Keselman A, Bender M, Carlson O, Egan J, Ladenheim B, Cadet JL, Becker KG, Wood W, Duffy K, Vinayakumar P, Maudsley S, Mattson MP (2007) Sex-dependent metabolic, neuroendocrine, and cognitive responses to dietary energy restriction and excess. Endocrinology 148: 4318–4333.
Matsunaga S, Yamada T, Mishima T, Sakamoto M, Sugiyama M, Wada M (2007) Effects of high-intensity training and acute exercise on in vitro function of rat sarcoplasmic reticulum. Eur J Appl Physiol 99: 641–649.
McCloskey DP, Adamo DS, Anderson BJ (2001) Exercise increases metabolic capacity in the motor cortex and striatum, but not in the hippocampus. Brain Res 891: 168–175.
Moriguchi S, Kato M, Sakai K, Yamamoto S, Shimizu E (1998) Exercise training restores decreased cellular immune functions in obese Zucker rats. J Appl Physiol 84: 311–317.
Murphy EA, Davis JM, Brown AS, Carmichael MD, Mayer EP, Ghaffar A (2004) Effects of moderate exercise and oat beta-glucan on lung tumor metastases and macrophage antitumor cytotoxicity. J Appl Physiol 97: 955–959.
Nickerson M, Elphick GF, Campisi J, Greenwood BN, Fleshner M (2005) Physical activity alters the brain Hsp72 and IL-1beta responses to peripheral E. coli challenge. Am J Physiol Regul Integr Comp Physiol 289: R1665–R1674.
Nieman DC (2007) Marathon training and immune function. Sports Med 37: 412–415.
Ogonovszky H, Sasvari M, Dosek A, Berkes I , Kaneko T , Tahara S , Nakamoto H , Goto S , Radak Z (2005) The effects of moderate, strenuous, and overtraining on oxidative stress markers and DNA repair in rat liver. Can J Appl Physiol 30: 186–195.
Ortega E, Giraldo E, Hinchado MD, Martín L, García JJ, De la Fuente M (2007) Neuroimmunomodulation during exercise: role of catecholamines as 'stress mediator' and/or 'danger signal' for the innate immune response. Neuroimmunomodulation 14: 206–212.
Pare WP, Houser VP (1973) Activity and food-restriction effects on gastric glandular lesions in the rat: the activity-stress ulcer. Bull Psychon Sci 2: 213–214.
Pieper DR, Ali HY, Benson LL, Shows MD, Lobocki CA, Subramanian MG (1995) Voluntary exercise increases gonadotropin secretion in male Golden hamsters. Am J Physiol 269: R179–R185.
Quadrilatero J, Hoffman-Goetz L (2005) Mouse thymocyte apoptosis and cell loss in response to exercise and antioxidant administration. Brain Behav Immun 19: 436–444.
Redila VA, Christie BR (2006) Exercise-induced changes in dendritic structure and complexity in the adult hippocampal dentate gyrus. Neuroscience 137: 1299–1307.
Rhea MR, Alderman BL (2004) A meta-analysis of periodized versus nonperiodized strength and power training programs. Res Q Exerc Sport 75: 413–422.
Roman V, van der Borght K, Leemburg SA, van der Zee E, Meerlo P (2005). Sleep restriction by forced activity reduces hippocampal cell proliferation. Brain Res 1065: 53–59.
Rosa EF, Freymüller E, Ihara SS, Aboulafia J, Nouailhetas VL (2008) Damaging effects of intense repetitive treadmill running on murine intestinal musculature. J Appl Physiol 104: 1410–1417.
Shima K, Zhu M, Noma Y, Mizuno A, Murakami T, Sano T, Kuwajima M (1997) Exercise training in Otsuka Long-Evans Tokushima Fatty rat, a model of spontaneous non–insulin-dependent

diabetes mellitus: effects on the B-cell mass, insulin content and fibrosis in the pancreas. Diabetes Res Clin Pract 35: 11–19.

Somani SM, Husain K (1996) Exercise training alters kinetics of antioxidant enzymes in rat tissues. Biochem Mol Biol Int 38: 587–595.

Steinacker JM, Lormes W, Reissnecker S, Liu Y (2004) New aspects of the hormone and cytokine response to training. Eur J Appl Physiol 91: 382–391.

Stranahan AM, Arumugam TV, Cutler RG, Lee K, Egan JM, Mattson MP (2008) Diabetes impairs hippocampal function through glucocorticoid-mediated effects on new and mature neurons. Nat Neurosci 11: 309–317.

Stranahan AM, Khalil D, Gould E (2006) Social isolation delays the positive effects of running on adult neurogenesis. Nat Neurosci 9: 526–533.

Stranahan AM, Khalil D, Gould E (2007) Running induces widespread structural alterations in the hippocampus and entorhinal cortex. Hippocampus 17: 1017–1022.

Stranahan AM, Lee K, Martin B, Maudsley S, Golden E, Cutter RG, Mattson MP (2009) Voluntary exercise and caloric restriction enhance hippocampal dendritic spine density and BDNF levels in diabetic mice. Hippocampus, DOI: 10. 1002/hipo. 20577.

Stranahan AM, Norman ED, Lee K, Cutler RG, Telljohann RS, Egan JM, Mattson MP (2008b) Diet-induced insulin resistance impairs hippocampal synaptic plasticity and cognition in middle-aged rats. Hippocampus 18: 1085–1088.

Swallow, JG, Garland T Jr, Carter PA, Zhan WZ, Sieck GC (1998) Effects of voluntary activity and genetic selection on aerobic capacity in house mice (*Mus domesticus*). J Appl Physiol 84: 69–76.

Trejo JL, Carro E, Torres-Aleman I (2001) Circulating IGF-1 mediates exercise-induced increases in the number of new neurons in the adult hippocampus. J Neurosci 21: 1628–1634.

Ueda H, Urano Y, Sakurai T, Kizaki T, Hitomi Y, Ohno H, Izawa T (2003) Enhanced expression of neuronal nitric oxide synthase in islets of exercise-trained rats. Biochem Biophys Res Commun 312: 794–800.

van Praag H, Kempermann G, Gage FH (1999) Running increases cell proliferation and neurogenesis in the adult mouse dentate gyrus. Nat Neurosci 2: 266–270.

Warren GL, Moran AL, Hogan HA, Lin AS, Guldberg RE, Lowe DA (2007) Voluntary run training but not estradiol deficiency alters the tibial bone–soleus muscle functional relationship in mice. Am J Physiol Regul Integr Comp Physiol 293: R2015–R2026.

Watson PA, Reusch JE, McCune SA, Leinwand LA, Luckey SW, Konhilas JP, Brown DA, Chicco AJ, Sparagna GC, Long CS, Moore RL (2007) Restoration of CREB function is linked to completion and stabilization of adaptive cardiac hypertrophy in response to exercise. Am J Physiol Heart Circ Physiol 293: H246–H259.

Zendzian-Piotrowska M, Górski J (1993) Metabolic adaptation to daily exercise of moderate intensity to exhaustion in the rat. Eur J Appl Physiol Occup Physiol 67: 77–82.

Zhao C, Teng EM, Summers RG Jr, Ming GL, Gage FH (2006) Distinct morphological stages of dentate granule neuron maturation in the adult mouse hippocampus. J Neurosci 26: 3–11.

Dietary Energy Intake, Hormesis, and Health

Bronwen Martin, Sunggoan Ji, Caitlin M. White, Stuart Maudsley, and Mark P. Mattson

Abstract The ability to adapt to varying levels of available energy in the form of food in the environment has allowed species to propagate and also thrive during times of energy surplus. However, in times when there is scant food available, similar evolutionary pressures have ensured that physiological systems can adapt to and utilize this food scarcity to their advantage. Considerable research has demonstrated that upon reduction of food intake, there are several beneficial effects upon cardiovascular, endocrinological, immune, and neuronal systems. Some of the effects of caloric restriction, however, tend to be exaggerated in many experimental cases due to biasing of overweight control subjects, yet reduction of total body weight still seems to engender beneficial effects for the individual. Some of the beneficial effects of caloric restriction are believed to arise from a reflexive response to the "stress" of reduced food intake. In conjunction with this is a similar hypothesis, known as "hormesis," which proposes in a similar vein that other forms of stress, such as toxicological stress, can also engender a "protective" set of physiological responses that shields the individual from further stresses. This chapter discusses how these two theories of protective responses—caloric restriction and hormesis—share many overlapping properties.

Keywords Caloric restriction · Energy homeostasis · Endocrinological · Neuroprotective · Adaptive · Evolutionary

Introduction

It is clear that many organisms have thrived throughout evolutionary history by developing mechanisms to control their physiology during times of either abundant or scarce food resources. Both mechanisms to store food as well as to extract the

---

B. Martin (✉)
Laboratory of Clinical Investigation, National Institutes of Health, National Institute on Aging, Baltimore, MD 21224, USA
e-mail: martinbro@mail.nih.gov

M.P. Mattson, E.J. Calabrese, *Hormesis*, DOI 10.1007/978-1-60761-495-1_7,

greatest amount of energy from available food have been developed. It has recently been proposed that not only do these mechanisms ensure short-term survival, but also that they may exert life-long effects by promoting a protective response, especially during times of stress. Indeed, even in the brief era of *Homo sapiens* evolution in multiple areas across the world, cultures have recognized the beneficial effects on health of limiting food intake for certain periods of time, either for religious reasons or during times of famine. The first widely recognized scientific study of the beneficial effects of caloric restriction (CR) was performed by McCay et al (1935). They demonstrated that rats fed with a diet containing indigestible cellulose possessed an extended mean and maximum lifespan compared to animals fed with a higher-caloric-density diet (McCay et al., 1935). Many studies since have confirmed this result and have extended it to other organisms, such as mice and fruitflies (Weindruch and Walford, 1988; Sprott, 1997; Chapman and Partridge, 1996; Houthoofd et al., 2002). It has been postulated that one of the broad mechanisms by which many of the beneficial effects of CR are created is through the invocation of cellular protection mechanisms. These general mechanisms are activated by the stressor and then serve to protect protein structure, energy production, and DNA stability for a considerable period of time that may even extend beyond the period of exposure to the stressor. In the specific case of CR, the stressor primarily takes the form of attenuated energy production. Therefore, it is possible that energetic processes are modified to utilize resources more efficiently, and critical systems are selectively protected against damage during this period. The hormetic hypothesis proposes that organisms can respond in a long-term protective manner to multiple environmental or physical stressors.

In this chapter, we use the paradigm of CR to explore the implications of the hormetic hypothesis for our understanding of how physiological mechanisms can adjust to and utilize factors that have previously been considered to be deleterious. We provide a comprehensive overview of many of the mechanisms by which CR can exert beneficial effects upon general physiology and how these may then be controlled by additional exogenous factors, such as xenobiotics or synthetically derived chemicals, to improve health and general well-being.

CR as a Hormetic Effector

Hormesis is defined as the beneficial effects that agents previously considered deleterious can have upon biological systems. These effects are ascribed to low doses of xenobiotic agents (e.g., natural defensive chemicals derived from plants or animals), environmental chemicals, or even ionizing radiation. The hormesis hypothesis proposes that biological and physiological systems can adjust to and thrive in the face of such insults as they induce a beneficial compensatory response in the organism. This response then readjusts the organism's physiology to make it more resistant to future stressors and utilize its endogenous resources in a more efficient manner. To this end, CR has often been considered an interesting paragon of this hypothesis because it involves the most basic stressor—direct inhibition of energetic processes. The lack of input energy may not be the only stress induced by CR, in that there may also be additional stresses, such as psychosocial stress and secondary cellular oxidative damage to proteins, lipids, or nucleic acids due to alterations in cellular metabolism. We will discuss how CR and hormesis have many overlapping properties and how the effects of CR on whole-body physiology could be considered hormetic. An overview of the overlapping properties of CR and hormesis is provided in Fig. 1.

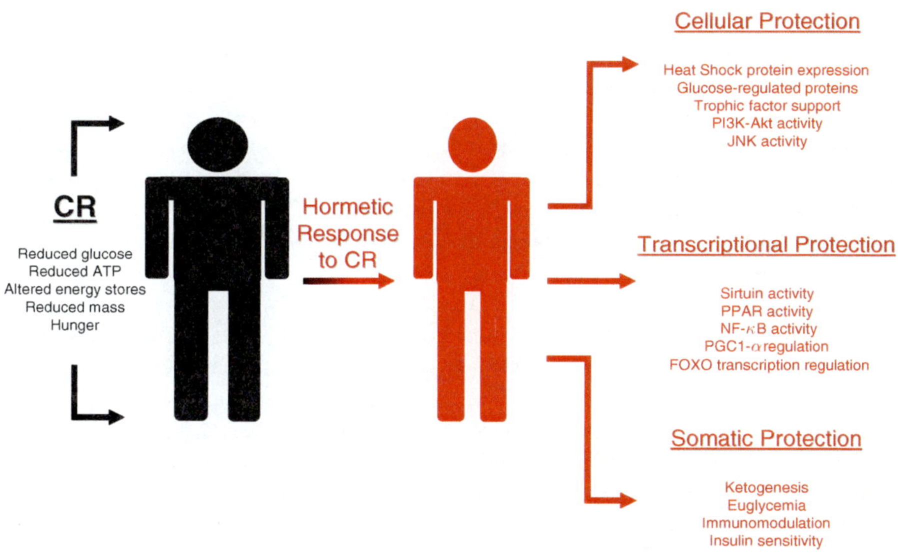

Fig. 1 Calorie restriction as a hormetic effector of multiple protective mechanisms. The induction of caloric restriction (CR) imposes a myriad of challenges to the body. To maintain survival and well-being in the face of the reduced capacity to produce and store energy for growth, reproduction, or homeostasis, the body responds to the CR state with a "hormetic-style" response. Hence, responsive mechanisms at multiple physiological levels are entrained to ameliorate and even employ the applied CR stress to the benefit of the body. These mechanisms occur at almost all levels of cellular and tissue organization—for example, with respect to maintenance of intermediary cell metabolism (cellular protection), generation of new proteins (transcriptional protection), and maintenance of whole-body endocrine/neurological axes (somatic protection)

CR and Cellular Stress Factors

In the face of reduced energy production (caused by reduced food intake), it is likely that highly energetic processes such as transcription and translation may be curtailed to conserve levels of nucleotide (adenosine or guanosine) triphosphates. On the other hand, stress-regulated proteins will be upregulated to exert cellular protective actions. For example, several different stress proteins have been measured in the brains from rats maintained on either ad libitum or CR diets for 3 months. Examples of such stress proteins include heat-shock proteins and glucose-regulated proteins. Heat-shock proteins comprise a huge family of distinct proteins that act as molecular chaperone proteins that interact with many different proteins in cells and function to ensure their proper folding, on one hand, and degradation of damaged proteins, on the other hand (Frydman, 2001; Gething, 1999). It has been shown that levels of some of these chaperone proteins are increased during the aging process as a protective response (Lee et al., 2000). Cell culture and in vivo studies have demonstrated the ability of heat-shock protein 70 (HSP-70) and glucose-regulated protein 78 (GRP-78) to be neuroprotective in experimental models of neurodegenerative disorders, excitotoxic stress, and oxidative injury (Lowenstein et al., 1991; Yu and Mattson, 1999; Warrick et al., 1999). Levels of HSP-70 and GRP-78 have also been found to be increased in the cortical, hippocampal, and striatal neurons of rats on a CR diet compared to age-matched ad libitum–fed animals (Lee et al., 1999; Mattson, 1998). It has also been demonstrated that heat-shock proteins can bind to and modify the activity of proapoptotic factors such as caspases (Beere et al., 2000; Ravagnan et al., 2001). These data may demonstrate that CR can induce a mild stress response in cells, presumably due to reduced energy—primarily glucose—availability. In addition to these cellular stress response mechanisms, it has been reported that reduced dietary energy results in increased levels of circulating corticosterone (Martin et al., 2006; 2007). Systemic corticosterone levels are usually associated positively with the stress state of the organism. In contrast to detrimental stressors such as chronic, uncontrollable stress (which can endanger cells through glucocorticoid receptor activation), reduced energy intake downregulates glucocorticoid receptors with maintenance of mineralocorticoid receptors in cells, which can act to prevent neuronal damage and death (Lee et al., 2000; Masoro, 2007). Thus, periods of energy scarcity may play a mechanistic role in triggering increases in cellular stress resistance and the repair of damaged proteins and cells.

CR Effects Upon Cytokine Levels

There is increasing evidence demonstrating the role of inflammatory mediators in the development of chronic, age-related disorders such as Alzheimer's disease. Pathophysiological activation of microglia is thought to be a major contributor to such conditions (Griffin, 2006). Any hormetic responsive mechanism—for example, to CR—may therefore also be able to promote healthy aging through effective amelioration of this chronic, uncontrolled immune response. It has been shown that circulating levels of interferon gamma (IFN-γ) are selectively elevated in monkeys maintained on a CR diet (Mascarucci et al., 2002). IFN- γ levels can also be enhanced in the hippocampus, where they can exert a profound excitoprotective action (Lee et al., 2006). Cytokine synthesis may also be affected by CR in peripheral tissue, as well as in the general circulation and in the central nervous system (CNS) (Bordone and Guarente, 2005). Tissue necrosis factor alpha (TNF-α) can be readily synthesized by adipose tissue, and upon its release it has been shown to affect insulin resistance and therefore long-term energy regulation and, potentially, eventual longevity (Feinstein et al., 1993). Of interest, the genetic removal of TNF-α receptors can improve insulin sensitivity, thus suggesting a potential mechanism of CR's hormetic actions, given that energy restriction can reduce the age-dependent upregulation of the TNF-α–controlling transcription factor NF-κB (Kim et al., 2000; Bordone and Guarente, 2005). Consistent with a role for suppression of NF-κB activity in the hormetic antiaging effect of CR, it was recently shown that genetic blockade of NF-κB for 2 weeks in the skin of chronologically aged mice restored the skin tissue characteristics back to those of young mice (Adler et al., 2007).

CR and Alterations in Neurotrophic Factors

A significant degree of attention has recently been paid to the emerging concept that neurological and endocrinological systems are functionally intertwined to a much greater extent than previously appreciated (Martin et al., 2008). Therefore, it is now accepted that alterations in neurological factors often can have dramatic effects upon peripheral physiology and vice versa. To this end, it has been demonstrated that there are often considerably strong effects of CR upon levels of neurotrophic agents such as brain-derived neurotrophic factor (BDNF). Given that CR is thought to induce a mild stress response in many cells in the periphery and the CNS, this can result in the activation of trophic hormone compensating mechanisms, for example, the upregulation of neurotrophic factors such as BDNF and glial cell line–derived neurotrophic factor (GDNF) (Bruce-Keller et al., 1999; Maswood et al., 2004). One of the primary neuroprotective mechanisms attributed to BDNF appears to be the BDNF-mediated activation of its cognate TrkB receptor tyrosine kinase. Activation of this receptor results in the potent stimulation of multiple signaling pathways associated with the ligand-dimerized TrkB receptor. Prominent among these TrkB signaling pathways is the phosphatidylinositol 3 kinase (PI3K)/protein kinase B (Akt) pathway that has been implicated in several of the CR protective mechanisms that will be discussed at greater length later in this chapter.

CR Effects Upon Glycemic Control

During episodic CR, the primary perturbation to the individual relates to the availability of glucose for oxidative respiration. The mechanisms by which energy is derived from alternate sources and how the remaining glucose is handled are crucial to the appreciation of the beneficial effects of CR paradigms. With respect to the hormetic effects of CR upon health and aging, it has been demonstrated that CR-induced reductions in glucose levels in the blood, integrated over time, can attenuate the levels of nonenzymatic glycation (a form of protein damage), as well as attenuate damaging oxyradical production (Weindruch and Sohal, 1997; Cefalu et al., 1995). Dietary energy restriction typically has the predictable effects upon somatic glycemic physiology with respect to levels of insulin and glucose, that is, CR causes a reduction in both. A longitudinal study on male rats (Masoro et al., 1992) showed that CR decreases the mean 24-hour plasma glucose concentration by about 15 mg/dL and the insulin concentration by about 50%. Of interest, however, CR animals utilized glucose at the same rate as ad libitum–fed animals despite the lower plasma glucose and markedly lower plasma insulin levels, indicating that their energy system was more efficient. CR has also been found to reduce plasma glucose and insulin concentrations and insulin sensitivity in fasting rhesus and cynomolgus monkeys (Kemnitz et al., 1994; Lane et al., 1996; Cefalu et al., 1997). The potential importance of modulating energy regulation by CR is the demonstration that loss-of-function mutations of the insulin signaling system result in life extension in three species: nematode worms (Kenyon et al., 1993), fruitflies (Clancy et al., 2001), and mice (Bluher et al., 2003). Therefore, it is highly likely that CR engenders complex physiological states that can result in enhanced glucose effectiveness, insulin responsiveness, or both, and that the maintenance of low levels of glucose and insulin may in part mediate the beneficial and life-extending hormetic actions of CR. It is likely that other hormetic agents could induce similar alterations in energy control and insulin sensitivity.

CR and Satiety/Adipose-Generated Hormones

Traditionally, the circulating hormones that control the desire and responsiveness of an organism towards food intake have largely been thought to serve only one function in the body. However, in recent years, our appreciation of hormones such as leptin and adiponectin has changed how we perceive the activities of these pluripotent hormones (Martin et al., 2007). Because CR can potently affect adiposity in most animals, it has significant effects upon the levels of satiety-related hormones synthesized by fat, for example, leptin and adiponectin (Meier and Gressner, 2004). As we have described, the primary hormetic action of CR could be mediated through its ability to subtly elevate an animal's stress response in a manner that engenders improved tolerance rather than excessive trauma. For example, CR regimens can effectively downregulate potentially damaging thyroid hormones via attenuation of circulating leptin levels (Barzilai and Gupta, 1999).

Adiponectin, on the other hand, has been shown to trigger increased insulin sensitivity (Meier and Gressner, 2004; Pajvani and Scherer, 2003) via upregulation of AMP-activated protein kinase (Wu et al., 2003). This kinase regulates glucose and fat metabolism in muscle in response to energy limitation (Musi et al., 2001) and has been shown to protect cells against metabolic stress (Culmsee et al., 2001). Adiponectin levels rise during CR, which suggests that this adipose-derived hormone might also have an important contributory role in the physiological shift to an enhanced insulin sensitivity and general protective responsivity in these animals (Combs et al., 2003). These data suggest that visceral adipose might be especially important in driving insulin sensitivity and potential pathogenesis (Bjorntorp, 1991), and, thus, alteration of this via CR may again impart a hormetic action upon whole-body energy regulation.

CR and Ketone Body Synthesis

Part of the beneficial response to CR appears to be a necessitated increase in the diversity of pathways by which the body can generate usable energy. CR and related paradigms have been shown to cause an increase in the somatic production of ketone bodies, for example, β-hydroxybutyrate. This simple ketone can be utilized by the body as an additional or alternative source of energy generation during periods of limited glucose availability (Vazquez et al., 1985; Mitchell et al., 1995). A CR-induced diversification to ketogenic energy generation not only may facilitate additional energy resources, but also may mediate a strong cytoprotective action. For example, rats fed a diet that favors the switch to in vivo production of ketones exhibit increased resistance to seizures (Bough et al., 1999). In addition, β-hydroxybutyrate can protect neurons in rodent models of neurodegenerative diseases and also reduces neurological damage incurred due to excitotoxicity during epileptic seizure activity (Kashiwaya et al., 2000; Gilbert et al., 2000).

CR and Sirtuin Activity

Genetic studies in yeast identified an important factor that seemed to control longevity—the silent information regulator 2 (SIR2), so denoted because it mediates a specific gene silencing action (Rine and Herskowitz, 1987). Mutagenesis of SIR2 that results in its inactivation shortens lifespan, and increased gene dose of SIR2 can conversely extend it (Kaeberlein et al., 1999). Given that dietary regulation has also shown to be a powerful modulator of both health and lifespan, it is reasonable to speculate that CR and SIR2 gene programs may converge to play an important role in these multiple and complex physiological pathways. The family of proteins encoded for by the mammalian SIR2 homolog (SIRT1) is collectively termed sirtuins. Several recent reports have shown increases in SIRT1 protein levels in response to food deprivation (Nemoto et al., 2004; Cohen et al., 2004).

Sirtuins act as NAD-dependent histone deacetylases (Imai et al., 2000; Landry et al., 2000). The mammalian SIRT1 enzyme, in addition to histones, can deacetylate many other substrates. For example, SIRT1 can deacetylate and downregulate NF-κB (Yeung et al., 2004). This action of CR upon NF-κB may therefore contribute to the ability of CR to increase insulin sensitivity and attenuate inflammatory mediators via modulation of TNF-α pathways. Therefore, sirtuins could be selectively regulated by the mild, controllable stress induced by CR and therefore play a role in hormetic response pathways to appropriate environmental cues.

In addition to a possible cytokine-dependent mechanism, sirtuins may mediate the effective role of adipose tissue in the physiological transference of the benefits of CR/hormetic regimes to the organism. Adipose tissue activity is strongly affected by the peroxisome proliferator-activated transcription factor receptor gamma (PPARγ) (Tontonoz et al., 1994). The PPARγ acts as a nuclear transcription factor that controls multiple gene targets connected to cell survival and responses to metabolic alterations. One such PPARγ target is the aP2 gene, which encodes a protein that assists fat storage. SIRT1 acts as a repressor of PPARγ, thereby downregulating genes such as the mouse aP2 gene (Picard et al., 2004). CR-induced SIRT1 activation subsequently can repress aP2 gene activity, causing an eventual promotion of fat mobilization into the blood to aid the organism's energy balance (Bordone and Guarente, 2005). Creation of additional energy supplies could aid the organism in the face of stressful environments that may engender hormetic physiological responses.

CR Modulation of PPARs and Cofactors

PPARs form functional heterodimers with retinoid X receptors (RXRs), and these heterodimers regulate the transcription of various genes involved in nutrient transport and metabolism, as well as in resistance to stress. There are three known subtypes of PPARs: α, δ, and γ. PPARs also recruit other proteins in addition to the RXR to mediate their full spectrum of activities—for example, PPARγ and coactivator 1 (PGC-1). PGC-1 has been shown to be regulated by dietary alteration and CR in both lower organisms and higher mammals. PGC-1 exists in two isoforms, α and β. PGC-1α interacts with the PPARγ to regulate brown fat differentiation during adaptation to environmental stresses such as cold temperature (Puigserver et al., 1998). During CR periods, when insulin levels are low, PGC-1α and PGC-1β gene expression is enhanced in rodents (Puigserver and Spiegelman, 2003; Herzig et al., 2001). PGC-1α and PGC-1β can coordinately regulate genes involved in gluconeogenesis and fatty acid β-oxidation in a number of organs during CR-induced moderate stress (Lin et al., 2002; Kamei et al., 2003; Kressler et al., 2002). In addition to PGC regulation during fasting, PPARα is also upregulated by fasting in liver, small and large intestines, thymus (Escher et al., 2001), and pancreas (Gremlich et al., 2005). During periods of fasting, PPARα knockout mice exhibit an inability to regulate genes involved in fatty acid β- and ω-oxidation and ketogenesis in the liver, kidney, and heart, along with lack of control of blood levels of glucose or ketone bodies (Kroetz et al., 1998; Sugden et al., 2001; Leone et al., 1999).

With specific respect to the liver, given that it contains significant glycogen, nutrient, and vitamin stores, one would expect a strong link between its functioning and potential hormetic effects of CR. Indeed, it has been demonstrated that CR can protect the liver from a wide range of environmental stressors, many of which induce damage through circulating inflammatory mediators (Kim et al., 2002; Bokov et al., 2004). In addition, PPARα activity has been shown to control liver responses to multiple forms of stress. For example, mice preexposed to PPARα agonists exhibit decreased cellular damage, increased tissue repair, and decreased mortality after exposure to both physical and chemical hepatic stressors such as thioacetamide (Anderson et al., 2002; Wheeler et al., 2003; Corton et al., 2004). Thus, PPARα may influence aging through the regulation of multiple damage and repair processes after exposure to a plethora of endogenous or environmental stressors. Indeed, even age-dependent reductions in the expression of PGC-1α have been reported (Ling et al., 2004), which may exacerbate the aging process by affecting available energy levels in cells. CR, however, has been shown to reverse this age-dependent decrease in PGC-1α, PPAR, and regulated genes (Weindruch et al., 2002; Kayo et al., 2001).

CR and Transcriptional Regulation

Many of the potential hormetic actions of CR may be exerted through profound changes in transcriptomes that control both damage response mechanisms and optimal energy utilization. One of the most important transcriptional mediators of this activity may be the FOXO family, which we will discuss further in this section. With respect to cellular degradation or survival in the face of stressful stimuli, one of the most important signaling processes is the canonical phosphatidylinositol 3-kinase (PI3K) and serine–threonine protein kinases (Akt-1/Akt-2/protein kinase B [PKB]) signaling cascade. As we have discussed, perhaps one of the most important adaptive responses to stressful inputs is the ability to modulate energy consumption. Therefore, it is not surprising that insulin is one of the most potent activators of the PI3K/Akt pathway in most species. For example, in the nematode worm, this pathway determines responses to aging and environmental stress (Guarente and Kenyon, 2000). Genetic manipulations of these worms that inhibit the insulin receptor-PI3K/Akt pathway increase the animal's lifespan, as well as its ability to withstand fluctuations in temperature and oxidative damage. These effects require reversal of negative regulation of the stress resistance factor Daf-16 (Libina et al., 2003). Daf-16 encodes a transcription factor containing a "forkhead" DNA-binding domain. Overexpression of Daf-16 in worms (Henderson and Johnson, 2001) significantly extends their lifespan. Daf-16 regulates the expression of an array of genes involved in xenobiotic metabolism and stress resistance (Murphy et al., 2003). Mammalian homologues of Daf-16 fall into the family of FoxO factors. Mammals possess four main groups of FOXO transcriptional controllers: FOXO1, FOXO3, FOXO4, and FOXO6. FOXO transcription factors belong to the larger Forkhead family of proteins, a family of transcriptional regulators characterized by the conserved "forkhead box" DNA-binding domain (Kaestner et al., 2000).

FOXO proteins control a wide array of genetic loci that are all linked by a common mechanism, in that they serve to control energy metabolism in the organism in response to environmental changes, such as food deprivation and cell stressors (Nakae et al., 2001; Yeagley et al., 2001; Kops et al., 2002; Tran et al., 2002).

In fact, FOXO transcription factors are tightly linked to mechanisms of energy regulation and their susceptibility to environmental alterations. Upon activation of insulin receptors, subsequent stimulation of the PI3K/Akt pathway results in phosphorylation of FOXOs in mammals. These phosphorylated factors then specifically bind 14-3-3 proteins, which facilitate their transport out of the nucleus, reducing their transcriptional activity. Thus, upon CR, there is a complex interplay between activation and inactivation of these FOXO factors. There are potentially beneficial effects of FOXO activation and inactivation, depending upon the prevailing cellular conditions. It has been demonstrated that CR can effectively "uncouple" insulin/IGF-I signaling to FOXO factors by markedly reducing plasma IGF-I and insulin levels in rats (Sonntag et al., 1999). CR-induced reductions in circulating insulin/IGF-I levels result in decreased Akt phosphorylation in liver (Al-Regaiey et al., 2005) and decreased PI3K expression in muscle (Argentino et al., 2005). Commensurate with these changes, CR also attenuates the expression of FOXO family members (Al-Regaiey et al., 2005; Tsuchiya et al., 2004). Therefore during CR, there are not only increases in nuclear/cytoplasmic FOXO ratios, but also in FOXO factor expression (Imae et al., 2003; Furuyama et al., 2003). The CR-mediated downregulation of insulin/IGF-I signaling that results in increases in the activity of FOXO factors critically regulates cell survival mechanisms and may be one of the prime loci of hormetically acting systemic agents. Thus FOXOs seem to exist at a nexus where cellular stress responses are connected to eventual survival mechanisms. For example, the stress-related protein kinase c-Jun N-terminal kinase 1 (JNK-1), which serves as a molecular sensor for various stressors, profoundly controls FOXO transcriptional action. In nematode worms, JNK-1 directly interacts with and phosphorylates the FOXO homologue Daf-16, and in response to heat stress, JNK-1 promotes the translocation of Daf-16 into the nucleus, while overexpression of JNK-1 in this organism leads to increases in lifespan and increased survival after heat stress (Oh et al., 2005). FOXO transcription factors seem to serve as molecular links between dietary modifications/stressful stimuli and eventual survival, thus making them important factors when considering the study of potential hormetic mechanisms. During CR, when the circulating levels of insulin/IGF-1 are attenuated to improve euglycemia, FOXO nuclear translocation results in the upregulation of a series of target genes that promote cell cycle arrest, stress resistance, and apoptosis. External stressful stimuli also trigger the relocalization of FOXO factors into the nucleus, thus allowing an adaptive response to stress stimuli. FOXO proteins translate environmental stimuli, including the stress induced by caloric restriction, into changes in gene expression patterns that may coordinate healthy aging and eventual longevity.

Conclusions

Typically, hormetic processes have been considered for chemical or even radiological stimuli, such as ionizing radiation and bioorganic pesticides. These stressors can place a multitude of pressures on an organism such that it may not survive the insult, or, conversely, it may adapt and modify its physiology to ameliorate or even use the pressure to aid its well-being. Removing or reducing an animal's primary source of energy— food—could be considered one of the most profound potential hormetic mechanisms. In this chapter, we have demonstrated that in response to reductions of food intake, many species, from worms to humans, can adapt, ameliorate, or even employ the stressful lack of food to their advantage. Many of these adaptive mechanisms seem to both control energy consumption and minimize cell damage caused by any imminent lack of energy for metabolic functions (see Fig.1 for summary). It is interesting to note that multiple species appear to possess similar mechanistic programs by which to deal with the CR-induced pressure, and it is highly likely that more-advanced species have been able to thrive only because of their inherited molecular mechanism of stress response from simpler organisms. By understanding the plethora of protective cellular mechanisms that are entrained by CR, we could use this valuable insight to elucidate the modus operandi of other novel hormetic factors or processes that may prove beneficial to our ongoing well-being.

References

Adler AS, Sinha S, Kawahara TL, Zhang JY, Segal E, Chang HY (2007) Motif module map reveals enforcement of aging by continual NF-kappaB activity. Genes Dev 21: 3244–3257.

Al-Regaiey KA, Masternak MM, Bonkowski M, Sun L, Bartke A (2005) Long-lived growth hormone receptor knockout mice: interaction of reduced IGF-1/insulin signaling and caloric restriction. Endocrinology 146: 851–860.

Anderson SP, Yoon L, Richard EB, Dunn CS, Cattley RC, Corton JC (2002) Delayed liver regeneration in peroxisome proliferator-activated receptor-alpha-null mice. Hepatology 36: 544–554.

Argentino DP, Dominici FP, Munoz MC, Al-Regaiey K, Bartke A, Turyn D (2005) Effects of long-term caloric restriction on glucose homeostasis and on the first steps of the insulin signaling system in skeletal muscle of normal and Ames dwarf (prop1(df)/prop1(df)) mice. Exp Gerontol 40: 27–35.

Barzilai N, Gupta G (1999) Revisiting the role of fat mass in the life extension induced by caloric restriction. J Gerontol A Biol Sci Med Sci 54: B89–B96.

Beere HM, Wolf BB, Cain K, et al. (2000) Heat-shock protein 70 inhibits apoptosis by preventing recruitment of procaspase-9 to the apaf-1 apoptosome. Nat Cell Biol 2: 469–475.

Bjorntorp P (1991) Metabolic implications of body fat distribution. Diab Care 14: 1132–1143.

Bluher M, Kahn BB, Kahn CR (2003) Extended longevity in mice lacking the insulin receptor in adipose tissue. Science 299: 572–574.

Bokov A, Chaudhuri A, Richardson A (2004) The role of oxidative damage and stress in aging. Mech Ageing Dev 125: 811–826.

Bordone L, Guarente L (2005) Calorie restriction, SIRT1 and metabolism: understanding longevity. Nat Rev Mol Cell Biol 6: 298–305.

Bough KJ, Valiyil R, Han FT et al. (1999) Seizure resistance is dependent upon age and calorie restriction in rats fed a ketogenic diet. Epilepsy Res 35: 21–28.

Bruce-Keller AJ, Umberger G, McFall R, et al. (1999) Food restriction reduces brain damage and improves behavioral outcome following excitotoxic and metabolic insults. Ann Neurol 45: 8–15.

Cefalu WT, Bell-Farrow AD, Wang ZQ, et al. (1995) Caloric restriction decreases age-dependent accumulation of the glycoxidation products, N epsilon-(carboxymethyl)-lysine and pentosidine, in rat skin collagen. J Gerontol A Biol Sci Med Sci 50: B337–B341.

Cefalu WT, Wagner JD, Wang ZQ, et al. (1997) A study of caloric restriction and cardiovascular aging in cynomolgus monkeys (*Macaca fascicularis*): a potential model for aging research. J Gerontol Biol Sci 52A: B10–B19.

Chapman T, Partridge L (1996) Female fitness in *Drosophila melanogaster* and interaction between the effect of nutrition and of encounter rate with males. Proc R Soc Lond B Biol Sci 263: 755–759.

Clancy DJ, Gems D, Harshman LG, et al. (2001) Extension of life span by loss of CHICO, a *Drosophila* insulin receptor substrate protein. Science 292: 104–106.

Cohen HY, Miller C, Bitterman KJ, et al. (2004) Calorie restriction promotes mammalian cell survival by inducing the SIRT1 deacetylase. Science 305: 390–392.

Combs TP, Berg AH, Rajala MW, et al. (2003) Sexual differentiation, pregnancy, calorie restriction, and aging affect the adipocyte-specific secretory protein adiponectin. Diabetes 52: 268–276.

Corton JC, Apte U, Anderson SP, et al. (2004) Mimetics of caloric restriction include agonists of lipid-activated nuclear receptors. J Biol Chem 279: 46204–46212.

Culmsee C, Monnig J, Kemp BE, et al. (2001) AMP-activated protein kinase is highly expressed in neurons in the developing rat brain and promotes neuronal survival following glucose deprivation. J Mol Neurosci 17: 45–58.

Escher P, Braissant O, Basu-Modak S, et al. (2001) Rat PPARs: quantitative analysis in adult rat tissues and regulation in fasting and refeeding. Endocrinology 142: 4195–4202.

Feinstein R, Kanety H, Papa MZ, et al. (1993) Tumor necrosis factor-alpha suppresses insulin-induced tyrosine phosphorylation of insulin receptor and its substrates. J Biol Chem 268: 26055–26058.

Frydman J (2001) Folding of newly translated proteins in vivo: the role of molecular chaperones. Annu Rev Biochem 70: 603–647.

Furuyama T, Kitayama K, Yamashita H, et al. (2003) Forkhead transcription factor FOXO1 (FKHR)-dependent induction of PDK4 gene expression in skeletal muscle during energy deprivation. Biochem J 375: 365–371.

Gething MJ (1999) Role and regulation of the ER chaperone BiP. Semin Cell Dev Biol 10: 465–472.

Gilbert DL, Pyzik PL, Freeman JM (2000) The ketogenic diet: seizure control correlates better with serum beta-hydroxybutyrate than with urine ketones. J Child Neurol 15: 787–790.

Gremlich S, Nolan C, Roduit R, et al. (2005) Pancreatic islet adaptation to fasting is dependent on peroxisome proliferator-activated receptor transcriptional up-regulation of fatty acid oxidation. Endocrinology 146: 375–382.

Griffin WS (2006) Inflammation and neurodegenerative diseases. Am J Clin Nutr 83: 470–474.

Guarente L, Kenyon C (2000) Genetic pathways that regulate ageing in model organisms. Nature 408: 255–262.

Henderson ST, Johnson TE (2001) Daf-16 integrates developmental and environmental inputs to mediate aging in the nematode *Caenorhabditis elegans*. Curr Biol 11: 1975–1980.

Herzig S, Long F, Jhala US, et al. (2001) CREB regulates hepatic gluconeogenesis through the coactivator PGC-1. Nature 413: 179–183.

Houthoofd K, Braeckman BP, Lenaerts I, et al. (2002) Axonic growth up-regulates mass-specific metabolic rate, stress resistance, and extends life-span in *Caenorhabditis elegans*. Exp Gerontol 37: 1371–1378.

Imae M, Fu Z, Yoshida A, et al. (2003) Nutritional and hormonal factors control the gene expression of FoxOS, the mammalian homologues of DAF-16. J Mol Catal Chem 30: 253–262.

Imai S, Armstrong CM, Kaeberlein M, et al. (2000) Transcriptional silencing and longevity protein sir2 is an NAD-dependent histone deacetylase. Nature 403:795–800.

Kaeberlein M, McVey M, Guarente L (1999) The SIR2/3/4 complex and SIR2 alone promote longevity in *Saccharomyces cerevisiae* by two different mechanisms. Genes Dev 13: 2570–2580.

Kaestner KH, Knochel W, Martinez DE (2000) Unified nomenclature for the winged helix/forkhead transcription factors. Genes Dev 14: 142–146.

Kamei Y, Ohizumi H, Fujitani Y, et al. (2003) PPARgamma coactivator 1beta/ERR ligand 1 is an ERR protein ligand, whose expression induces a high-energy expenditure and antagonizes obesity. Proc Natl Acad Sci USA 100: 12378–12383.

Kashiwaya Y, Takeshima T, Mori N, et al. (2000) D-beta-hydroxybutyrate protects neurons in models of Alzheimer's and Parkinson's disease. Proc Natl Acad Sci USA 97: 5440–5444.

Kayo T, Allison DB, Weindruch R, et al., (2001) Influences of aging and caloric restriction on the transcriptional profile of skeletal muscle from rhesus monkeys. Proc Natl Acad Sci USA 98: 5093–5098.

Kemnitz W, Roecker EB, Weindruch R, et al. (1994) Dietary restriction increases insulin sensitivity and lowers blood glucose in rhesus monkeys. Am J Physiol 266: E540–E547.

Kenyon C, Chang J, Gensch A, et al. (1993) A *C. elegans* mutant that lives twice as long as wild type. Nature 366: 461–464.

Kim HJ, Jung KJ, Yu BP, et al. (2002) Modulation of redox-sensitive transcription factors by calorie restriction during aging. Mech Ageing Dev 123: 1589–1595.

Kim HJ, Kim KW, Yu BP, et al. (2000) The effect of age on cyclooxygenase-2 gene expression: NF-κB activation and IκBα degradation. Free Radic Biol Med 28: 683–692.

Kops GJ, Dansen TB, Polderman PE, et al. (2002) Forkhead transcription factor FOXO3a protects quiescent cells from oxidative stress. Nature 419: 316–321.

Kressler D, Schreiber SN, Knutti D, et al. (2002) The PGC-1-related protein PERC is a selective coactivator of estrogen receptor alpha. J Biol Chem 277: 13918–13925.

Kroetz DL, Yook P, Costet P, et al. (1998) Peroxisome proliferator-activated receptor alpha controls the hepatic CYP4A induction adaptive response to starvation and diabetes. J Biol Chem 273: 31581–31589.

Landry J, Sutton A, Tafrov ST, et al. (2000) The silencing protein SIR2 and its homologs are NAD-dependent protein deacetylases. Proc Natl Acad Sci USA 97: 5807–5811.

Lane MA, Baer DJ, Rumpler WV, et al. (1996) Calorie restriction lowers body temperature in rhesus monkeys, consistent with a postulated anti-aging mechanism in rodents. Proc Natl Acad Sci USA 93: 4159–4164.

Lee J, Bruce-Keller AJ, Kruman Y, et al. (1999) 2-Deoxy-d-glucose protects hippocampal neurons against excitotoxic and oxidative injury: evidence for the involvement of stress proteins. J Neurosci Nurs 57: 48–61.

Lee J, Kim SJ, Son TG, et al. (2006) Interferon-gamma is up-regulated in the hippocampus in response to intermittent fasting and protects hippocampal neurons against excitotoxicity. J Neurosci Nurs 83: 1552–1557.

Lee CK, Weindruch R, Prolla TA (2000) Gene-expression profile of the ageing brain in mice. Nat Genet 25: 294–297.

Leone TC, Weinheimer CJ, Kelly DP (1999) A critical role for the peroxisome proliferator-activated receptor alpha (PPARalpha) in the cellular fasting response: the PPARalpha-null mouse as a model of fatty acid oxidation disorders. Proc Natl Acad Sci USA 96: 7473–7478.

Libina N, Berman JR, Kenyon C (2003) Tissue-specific activities of *C. elegans* DAF-16 in the regulation of lifespan. Cell 115: 489–502.

Lin J, Puigserver P, Donovan J, et al. (2002) Peroxisome proliferator-activated receptor gamma coactivator 1beta (PGC-1beta), a novel PGC-1-related transcription coactivator associated with host cell factor. J Biol Chem 277: 1645–1648.

Ling C, Poulsen P, Carlsson E, et al. (2004) Multiple environmental and genetic factors influence skeletal muscle PGC-1alpha and PGC-1beta gene expression in twins. J Clin Invest 114: 1518–1526.

Lowenstein DH, Chan PH, Miles MF (1991) The stress protein response in cultured neurons: characterization and evidence for a protective role in excitotoxicity. Neuron 7: 1053–1060.

Martin B, Golden E, Keselman A, et al. (2008) Therapeutic perspectives for the treatment of Huntington's disease: treating the whole body. Histol Histopathol 23: 237–250.

Martin B, Mattson MP, Maudsley S (2006) Caloric restriction and intermittent fasting: two potential diets for successful brain aging. Ageing Res Rev 5: 332–353.

Martin B, Pearson M, Kebejian L, et al. (2007) Sex-dependent metabolic, neuroendocrine, and cognitive responses to dietary energy restriction and excess. Endocrinology 148: 4318–4333.

Mascarucci P, Taub D, Saccani S, et al. (2002) Cytokine responses in young and old rhesus monkeys: effect of caloric restriction. J Interferon Cytokine Res 22: 565–571.

Masoro EJ (2007) The role of hormesis in life extension by dietary restriction. Interdiscip Top Gerontol 35: 1–7.

Masoro EJ, McCarter RJM, Katz MS, et al. (1992) Dietary restriction alters the characteristics of glucose fuel use. J Gerontol Biol Sci 47: B202–B208.

Maswood N, Young J, Tilmont E, et al. (2004) Caloric restriction increases neurotrophic factor levels and attenuates neurochemical and behavioral deficits in a primate model of Parkinson's disease. Proc Natl Acad Sci USA 101: 18171–18176.

Mattson MP (1998) Experimental models of Alzheimer's disease. Sci Med 7: 16–25.

McCay CM, Crowell MF, Maynard LA (1935) The effect of retarded growth upon the length of life-span and upon the ultimate body size. J Nurse Pract 10: 63–79.

Meier U, Gressner AM (2004) Endocrine regulation of energy metabolism: review of pathobiochemical and clinical chemical aspects of leptin, ghrelin, adiponectin, and resistin. Clin Chem 50: 1511–1525.

Mitchell GA, Kassovska-Bratinova S, Boukaftane Y, et al. (1995) Medical aspects of ketone body metabolism. Clin Invest Med 18: 193–216.

Murphy CT, McCarroll SA, Bargmann CI, et al. (2003) Genes that act downstream of DAF-16 to influence the lifespan of *Caenorhabditis elegans*. Nature 424: 277–283.

Musi N, Hayashi T, Fujii N, et al. (2001) AMP-activated protein kinase activity and glucose uptake in rat skeletal muscle. Am J Physiol Endocrinol Metab 280: 677–684.

Nakae J, Kitamura T, Silver DL, et al. (2001) The forkhead transcription factor foxo1 (fkhr) confers insulin sensitivity onto glucose-6-phosphatase expression. J Clin Invest 108: 1359–1367.

Nemoto S, Fergusson MM, Finkel T (2004) Nutrient availability regulates SIRT1 through a forkhead-dependent pathway. Science 306: 2105–2108.

Oh SW, Mukhopadhyay A, Svrzikapa N, et al. (2005) JNK regulates lifespan in *Caenorhabditis elegans* by modulating nuclear translocation of forkhead transcription factor/DAF-16. Proc Natl Acad Sci USA 102: 4494–4499.

Pajvani UB, Scherer PE (2003) Adiponectin: systemic contributor to insulin sensitivity. Curr Diab Rep 3: 207–213.

Picard F, Kurtev M, Chung N, et al. (2004) Sirt1 promotes fat mobilization in white adipocytes by repressing PPAR-gamma. Nature 429: 771–776.

Puigserver P, Spiegelman BM (2003) Peroxisome proliferator-activated receptor-gamma coactivator 1 alpha (PGC-1 alpha): transcriptional coactivator and metabolic regulator. Endocr Rev 24: 78–90.

Puigserver P, Wu Z, Park CW, et al. (1998) A cold-inducible coactivator of nuclear receptors linked to adaptive thermogenesis. Cell 92: 829–839.

Ravagnan L, Gurbuxani S, Susin SA, et al. (2001) Heat-shock protein 70 antagonizes apoptosis-inducing factor. Nat Cell Biol 3: 839–843.

Rine J, Herskowitz I (1987) Four genes responsible for a position effect on expression from HML ANS HMR in *Saccharomyces cerevisiae*. Genetics 116: 9–22.

Sonntag WE, Lynch CD, Cefalu WT, et al. (1999) Pleiotropic effects of growth hormone and insulin-like growth factor (IGF)-1 on biological aging: inferences from moderate caloric-restricted animals. J Gerontol Biol Sci 54: 521–538.

Sprott RL (1997) Diet and calorie restriction. Exp Gerontol 32: 205–214.

Sugden MC, Bulmer K, Gibbons GF, et al. (2001) Role of peroxisome proliferator-activated receptor-alpha in the mechanism underlying changes in renal pyruvate dehydrogenase kinase isoform 4 protein expression in starvation and after refeeding. Arch Biochem Biophys 395: 246–252.

Tontonoz P, Hu E, Graves RA, et al. (1994) MPPARγ 2: tissue-specific regulator of an adipocyte enhancer. Genes Dev 8: 1224–1234.
Tran H, Brunet A, Grenier JM, et al. (2002) DNA repair pathway stimulated by the forkhead transcription factor FOXO3a through the gadd45 protein. Science 296: 530–534.
Tsuchiya T, Dhahbi JM, Cui X, et al. (2004) Additive regulation of hepatic gene expression by dwarfism and caloric restriction. Physiol Genom 17: 307–315.
Vazquez JA, Morse EL, Adibi SA (1985) Effect of dietary fat, carbohydrate, and protein on branched-chain amino acid catabolism during caloric restriction. J Clin Invest 76: 737–743.
Warrick JM, Chan HY, Gray-Board GL, et al. (1999) Suppression of polyglutamine-mediated neurodegeneration in *Drosophila* by the molecular chaperone HSP70. Nat Genet 23: 425–428.
Weindruch R, Kayo T, Lee CK, et al. (2002) Gene expression profiling of aging using DNA microarrays. Mech Ageing Dev 123: 177–193.
Weindruch R, Sohal RS (1997) Seminars in Medicine of the Beth Israel Deaconess Medical Center: caloric intake and aging. N Engl J Med 337: 986–994.
Weindruch R, Walford RL (1988) The retardation of aging and disease by dietary restriction. Springfield, IL: Charles C Thomas.
Wheeler MD, Smutney OM, Check JF, et al. (2003) Impaired ras membrane association and activation in PPARalpha knockout mice after partial hepatectomy. Am J Physiol Gastrointest Liver Physiol 284: G302–G312.
Wu X, Motoshima H, Mahadev K, et al. (2003) Involvement of AMP-activated protein kinase in glucose uptake stimulated by the globular domain of adiponectin in primary rat adipocytes. Diabetes 52: 207–213.
Yeagley D, Guo S, Unterman T, et al. (2001) Gene- and activation-specific mechanisms for insulin inhibition of basal and glucocorticoid-induced insulin-like growth factor binding protein-1 and phosphoenolpyruvate carboxykinase transcription. Roles of forkhead and insulin response sequences. J Biol Chem 276: 33705–33710.
Yeung F, Hoberg JE, Ramsey CS, et al. (2004) Modulation of NF-kappaB–dependent transcription and cell survival by the SIRT1 deacetylase. EMBO J 23: 2369–2380.
Yu ZF, Mattson MP (1999) Dietary restriction and 2-deoxyglucose administration reduce focal ischemic brain damage and improve behavioral outcome: evidence for a preconditioning mechanism. J Neurosci Nurs 57: 830–839.

Couch Potato: The Antithesis of Hormesis

Mark P. Mattson, Alexis Stranahan, and Bronwen Martin

Abstract The impact of hormesis on health can be further appreciated by consideration of the "couch potato" lifestyle. When cells in the body and brain are not challenged, they become complacent and are therefore vulnerable to injury and disease. Lack of physical and mental exercise, in combination with excessive food intake, results in a condition called insulin resistance that is a harbinger of diabetes and cardiovascular disease. On the other hand, when fewer calories are consumed and when more energy is expended (exercise), cells are subjected to a mild metabolic stress. They respond to this mild stress adaptively by increasing their ability to take up glucose in respond to insulin. This hormesis response is, in part, responsible for the ability of dietary energy restriction and exercise to ward off diabetes and cardiovascular disease. However, obesity and diabetes are not the only adverse physiological consequences of being a "couch potato." Exercise and dietary energy restriction improve the functional efficiency of the heart and gut through a hormetic mechanism that involves increased activity of the parasympathetic component of the autonomic nervous system. As a consequence, heart rate and blood pressure are decreased and gut motility is increased, thereby reducing the risk of heart disease, stroke, and colon cancer. Relatively underappreciated is the contribution of the lack of mental challenges to the poor health associated with the couch potato lifestyle. Individuals who engage in intellectually challenging occupations or hobbies may be at reduced risk for Alzheimer's disease because of the beneficial stress imposed on the neurons when they are challenged. Studies have shown that neurons respond to mental and physical activity by increasing their production of "neurotrophic factors" that may help them to resist disorders such as Alzheimer's disease and Parkinson's disease.

Keywords Calories · Diabetes · Exercise · Food addiction · Insulin resistance · Obesity · Sedentary lifestyle

---

M.P. Mattson (✉)
Laboratory of Neurosciences, National Institutes of Health, National Institute on Aging, Baltimore, MD 21224, USA
e-mail: mattsonm@grc.nia.nih.gov

M.P. Mattson, E.J. Calabrese, *Hormesis*, DOI 10.1007/978-1-60761-495-1_8,

The "Couch Potato" Caricature

The extreme example of an unhealthy lifestyle is the person who gets no exercise, eats excessive amounts of high-calorie foods, and avoids intellectually challenging tasks. The stereotypical features of such a "couch potato" include an obese person reclining on a couch, watching television, drinking beer or sodas, and eating potato chips, pizza or French fries (Fig.1). The couch potato lifestyle and its associated morbidity and mortality has only recently become common in modern societies during the last two or three generations (CDC. Diabetes, 2008). Couch potatoes were rare prior to the 20th century and are also currently rare in nonindustrialized countries. A sedentary lifestyle (Eggermont et al., 2006; Hamilton et al., 2007), excessive calorie intake (Luchsinger et al., 2002; Astrup et al., 2008), and cognitive complacency (Stern, 2006) have all been shown to increase the risk of developing one or more major diseases, including diabetes, cardiovascular disease, cancers, and Alzheimer's disease. Thus, the couch potato subjects himself or herself to a triple whammy that accelerates the aging process and renders cells throughout the body and brain vulnerable to dysfunction and disease, inevitably resulting in premature death. The question then becomes how, at the cellular and molecular levels, does a couch potato lifestyle promote disease processes? In this chapter we propose that the fundamental problem is that cells in the body of a couch potato are not subjected to the kinds of mild stress that activate adaptive stress response pathways. The cells therefore become complacent, unaware that they are thereby endangering the health and survival of the individual in which they reside.

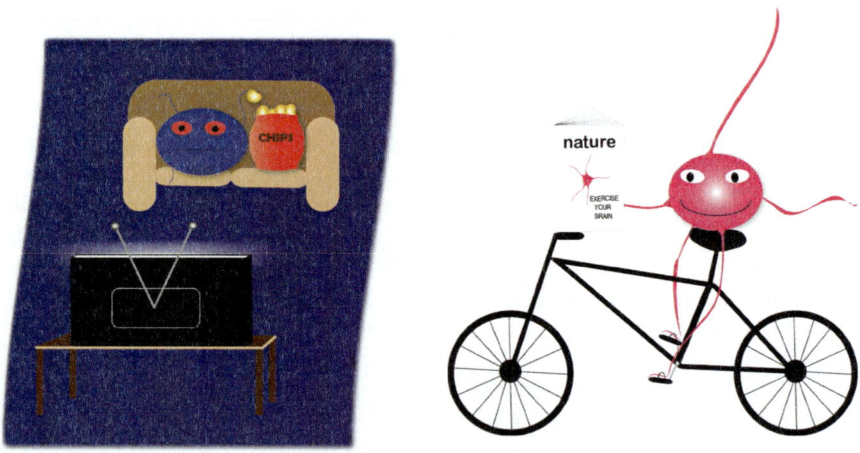

Fig. 1 The "couch potato" neuron versus the active neuron. The "neuron" on the left represents the effects of an unhealthy "couch potato" lifestyle of engaging in no exercise, eating excessive amounts of high calorie foods, and avoiding intellectually challenging tasks. The couch potato neuron appears unhealthy, in contrast to the active neuron on the right that is engaging in physically and mentally stimulating activity

Advances in Technology Reveal the Dangers of a Sedentary Lifestyle

What do telephones, automobiles, planes, trains, televisions, and computers all have in common? They all engender a sedentary lifestyle. There is solid evidence that the sharp increase in the incidence of obesity and diabetes is, in part, the result of the development of technologies that reduce the need to walk (Nestle and Jacobson, 2000; Lajunen et al., 2007; Shields and Tremblay, 2008). Thus, many of us drive back and forth to work, sit in front of a computer most of the day—communicating with coworkers by email instead of walking down the hall or up the stairs to speak with them in person—and watch television or surf the Internet for entertainment instead of playing or working outside. Technological advances have also resulted in a marked decrease in the number of individuals whose occupations are physically demanding. For example, someone growing up on a farm 40 or 50 years ago would have stacked hay bales on a wagon behind a baler and then restacked the bales in the barn, a considerable amount of work. Today, baling machines allow one worker to make bales and transfer them into a barn or pasture without ever leaving his or her tractor. Similar advances in technology have lessened or eliminated physical activities for many other occupations, including those in construction, landscaping, and even newspaper delivery. Even "sporting" activities are becoming less physically challenging; for example, most golfers now use golf carts.

Because the shift from active to inactive lifestyles has occurred rapidly, there are as yet no adverse effects of a sedentary lifestyle on reproductive fitness. In fact, a sedentary lifestyle combined with excessive dietary energy intake has likely been a major factor in the decrease in the age of puberty onset, particularly in females (Bau et al., 2009). It is therefore unlikely that genes that may protect against obesity will be selected for in the future even as childhood obesity reaches epidemic proportions. On the other hand, advances in understanding the neurobiological and metabolic underpinnings of obesity will likely result in the development of drugs that mimic the effects of exercise on cells. Measures to increase physical activity in the workplace might include policies that require nonhandicapped employees to use stairways, to spend 30 minutes exercising during the workday (walking or using exercise facilities), and so on. It will also be important to develop technologies that incorporate exercise into the daily routine. For example, many of us spend several hours each day working in front of a computer. Workstations are being designed such that one can exercise at light to moderate intensity levels while reading and typing (McAlpine et al., 2007).

Cellular and Molecular Mechanisms of Exercise Hormesis

The effects of exercise on various organ systems are reviewed in the chapter Exercise-Induced Hormesis. We (Stranahan et al., 2008) and others (Gomez-Pinilla, 2008; Radak et al., 2008) believe that the major reason that lack of exercise is bad

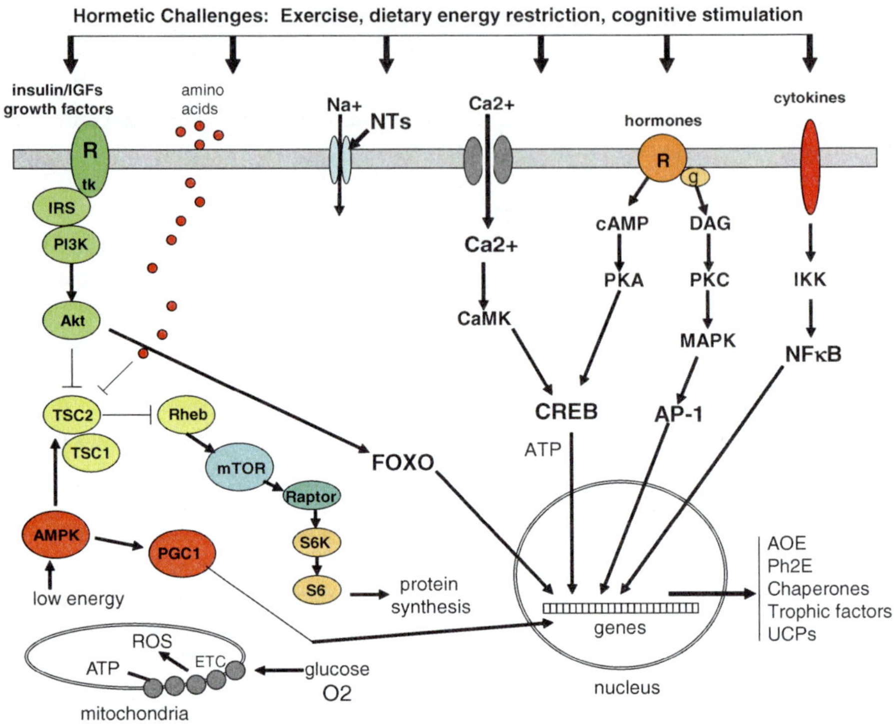

Fig. 2 Adaptive cellular stress response pathways tapped by exercise, dietary energy restriction, and cognitive stimulation, which are untapped in the cells of couch potatoes. In response to mild stress (oxidative, metabolic, ionic, etc.), one or more adaptive stress response pathways are activated in cells. For example, in response to energetic stress (fasting or exercise), insulin and insulin-like growth factor (IGF) activate receptors (R) with intrinsic tyrosine kinase (tk) activity. The activated receptor phosphorylates the insulin receptor substrate protein (IRS), thus engaging a kinase cascade involving phosphatidylinositol 3 kinase (PI3K) and Akt kinase. Two Akt substrates that are important in adaptive responses of cells to stress are the proteins FOXO and TSC2. FOXO is a transcription factor that, when activated, binds to regulatory elements of genes that encode cytoprotective proteins such as the antioxidant enzyme (AOE) manganese superoxide dismutase. Phosphorylation of TSC2 by Akt inhibits TSC2, resulting in the derepression of Rheb and the activation of mTOR (mammalian target of rapamycin). mTOR then activates Raptor, which, in turn, activates S6 kinase (S6K), thus activating S6, a protein critical for protein synthesis. When cellular energy levels are low, AMP kinase (AMPK) is activated, resulting in the activation of TSC2 and therefore inhibition of the mTOR signaling pathway. In addition, AMPK activates PGC1-α, a transcriptional regulator that increases the expression of genes that encode proteins involved in cellular energy metabolism and stress resistance. Because they produce ATP as well as reactive oxygen species (ROS), mitochondria (*lower left*) play pivotal roles in many cellular stress responses. Exercise, dietary energy restriction, and cognitive stimulation may induce the production of certain hormones (e.g., epinephrine and GLP-1) and cytokines (e.g., TNF) that activate adaptive stress response pathways. For example, hormone receptors coupled to GTP-binding proteins (g) can induce the production of cAMP or diacylglycerol (DAG); cAMP activates protein kinase A (PKA), while DAG activates protein kinase C (PKC). PKA, together with calcium/calmodulin-dependent protein kinase (CaMK), activates the transcription factor CREB (cyclic AMP response element–binding protein). PKC activates mitogen-activated protein kinases (MAPK), which, in turn, activate

for health is that exercise challenges and stimulates cells throughout the body so that they engage signaling pathways that enhance the ability of the cells to cope with stress and resist disease. This "use it or lose it" concept fits nicely under the umbrella of hormesis: When cells are challenged, they respond adaptively. Exercise involves highly coordinated vigorous activity in nerve cells and muscle cells (both skeletal muscle cells and cardiac muscle cells). This electrochemical activity increases cellular energy (ATP and NAD^+) levels to maintain the function of ion-motive ATPases in neurons and muscle cells and to fuel the contractile apparatus in muscle cells. The increased mitochondrial electron transport chain activity required to generate the additional ATP also results in the production of oxygen free radicals. The influx of Ca^{2+} and its release from the endoplasmic reticulum creates additional stress. These different subcellular stressors activate one or more hormetic signaling pathways. For example, the increased ATP consumption and Ca^{2+} influx that occurs in neurons and muscle cells during exercise stimulates AMP-activated protein kinase (AMPK), which, in turn induces mitochondrial ATP production while simultaneously inhibiting cell growth and proliferation (Towler and Hardie, 2007). One target of AMPK is the mammalian target of rapamycin (mTOR), a protein that stimulates cell growth and division. AMPK inhibits mTOR, thereby preserving cellular energy reserves (Fig. 2).

PGC1α (peroxisome-proliferator–activated receptor-γ coactivator 1α) is a transcriptional coactivator that is stimulated by exercise and regulates the expression of genes encoding proteins involved in cellular glucose and lipid metabolism (Benton et al., 2008; Handschin and Spiegelman, 2008). The PGC1α pathway is downregulated by sedentary lifestyles and in obesity and diabetes. In response to cellular energy depletion, as occurs during exercise, AMPK phosphorylates PGC1α, thereby increasing its activity. Muscle cells in transgenic mice that overexpress PGC1α exhibit improved fatty acid oxidation and insulin-induced glucose transport and are more resistant to fatigue and exhibit greater endurance than nontransgenic mice (Lin et al., 2002; Calvo et al., 2008). On the other hand, mice that lack PGC1α in skeletal muscle exhibit more fast-twitch muscle fibers and become exhausted more easily than wild-type mice during exercise (Handschin et al., 2007).

Peptide hormones produced by gut or adipose cells play major roles in mediating adaptive responses of animals and humans to changes in energy intake and expenditure. Alterations in the production of these hormones and/or their actions on target cells play roles in the diseases that couch potatoes often develop. These hormones include leptin, ghrelin, glucagon-like peptide 1 (GLP-1), adiponectin, and resistin (Baggio and Drucker, 2006; Doyle and Egan, 2007; Sun et al., 2007; Antuna-Puente

Fig. 2 (continued) the transcription factor AP-1. Activation of some cytokine receptors results in the activation of the transcription factor NF-kB. Some of the proteins that mediate the beneficial effects of mild stress on cells are listed at the lower right and include AOEs, phase 2 enzymes (Ph2E), protein chaperones such as heat-shock proteins, trophic factors such as IGFs and BDNF (brain-derived neurotrophic factor), and mitochondrial uncoupling proteins (UCPs)

et al., 2008). The two major hormetic manipulations of energy metabolism—dietary energy restriction and exercise—decrease levels of leptin and insulin and increase target cell sensitivity to these hormones while increasing levels of adiponectin and ghrelin (Dyck, 2005; Martin et al., 2007; Shinmura et al., 2007; Vu et al., 2007). Increasing evidence suggests that adiponectin and ghrelin have beneficial effects on cells in many different tissues, including those in the cardiovascular (García and Korbonits, 2006; Ouchi et al., 2006), immune (Dixit and Taub, 2005; Tilg and Wolf, 2005), and nervous (Diano et al., 2006; Hoyda et al., 2009) systems. GLP-1, in addition to increasing insulin sensitivity and insulin production, can protect the heart against ischemic injury (Bose et al., 2005) and can protect neurons against dysfunction and degeneration in animal models of epileptic seizures, Huntington's disease, Parkinson's disease, stroke and peripheral neuropathy (During et al., 2003; Perry et al., 2007; Li et al., 2009; Martin et al., 2009).

The hormones that mediate beneficial effects of exercise exert their effects on cells by activating cell surface receptors coupled to intracellular signaling cascades that induce the expression of cytoprotective proteins (Fig. 2). When insulin binds to its receptor, the intracellular tyrosine kinase domains of the receptor become activated, resulting in autophosphorylation of the receptor and recruitment of IRS-1 to the receptor. IRS-1 then activates PI3 kinase, which, in turn, activates Akt kinase. Akt phosphorylates and thereby inhibits GSK-3β, resulting in increased glycogenesis. Akt also phosphorylates, and thereby inhibits, the nuclear localization of FOXO transcription factors (Greer and Brunet, 2005). Upon binding of leptin to its receptor (OB-Rb), JAK2 associates with the cytoplasmic portion of the OB-Rb and phosphorylates several tyrosine residues on OB-Rb. The protein STAT-3 is then recruited to phosphorylated Ob-Rb, and STAT-3 is then phosphorylated by JAK2, resulting in the formation of STAT-3 dimers, which then translocate to the nucleus and regulate gene transcription (Fruhbeck, 2006). GLP-1 receptors are coupled to the membrane-associated GTP-binding protein Gs, which, in turn, activates adenylate cyclase, resulting in the production of cyclic AMP. Cyclic AMP activates protein kinase A, which then activates the transcription factor cyclic AMP response element–binding protein (CREB). Examples of target genes of the insulin, leptin, and GLP-1 signaling pathways include manganese superoxide dismutase, Bcl-2 family members, and growth factors.

Calories In, Disease Out

Excessive energy intake promotes disease and shortens lifespan, and, conversely, dietary energy restriction protects against disease and lengthens life (Mattson and Wan, 2005; Fontana and Klein, 2007). Studies of rats and mice have shown that dietary energy restriction lengthens lifespan and can counteract disease processes in models of cardiovascular disease (Ahmet et al., 2005; Mager et al., 2006), cancers (Hursting et al., 2003), and neurodegenerative disorders such as Alzheimer's,

Parkinson's, and Huntington's diseases (Duan and Mattson, 1999; Duan et al., 2003; Halagappa et al., 2007). In such rodent studies, the control groups are fed ad libitum and are relatively sedentary, and so those findings should be interpreted as showing that energy restriction improves health and increases longevity in couch potato rats and mice. However, studies of normal-weight nonhuman primates do suggest that dietary energy restriction is beneficial for non–couch potatoes as well. For example, in normal weight rhesus monkeys, moderate caloric restriction improved a range of health indicators (Mattison et al., 2007) and protected dopaminergic neurons in a model of Parkinson's disease (Maswood et al., 2004). Studies of humans have shown that overeating promotes type 2 diabetes, cardiovascular disease, and some cancers (Everitt and Le Couteur, 2007). Controlled trials of caloric restriction in humans support the data from studies of animals by showing improvements in a range of health indicators (Redman et al., 2008) and by demonstrating beneficial effects in patients with chronic diseases such as asthma (Johnson et al., 2007).

A high-calorie diet is detrimental for health for at least two major reasons. First, many cells in someone who overeats are subjected to a higher steady-state amount of oxygen free radicals because of increased mitochondrial oxidative phosphorylation resulting from increased glucose availability. As a consequence, levels of oxidatively modified proteins, lipids, and DNA are increased (Vincent et al., 2007). Second, dietary energy excess is unhealthy because it promotes "complacency" in cells, thereby rendering them unprepared to cope with stress and resist disease. Thus, a low-calorie diet and intermittent fasting impose a mild stress, to which cells respond adaptively by increasing the production of protein chaperones, mitochondrial uncoupling proteins, phase 2, proteins and antioxidant enzymes (Masoro, 2005). In addition, dietary energy restriction can induce the expression of growth factors that promote cell survival and enhance functional plasticity. For example, intermittent fasting increases the production of brain-derived neurotrophic factor (BDNF) in neurons and may thereby enhance learning and memory and protect neurons against age-related and injury-induced degeneration (Arumugam et al., 2006). Hormesis therefore plays an important role in the beneficial effects of dietary energy restriction, and, conversely, hormetic pathways are not activated in the cells of couch potatoes.

In One Ear and Out the Other

Epidemiological data suggest that individuals who engage in intellectually challenging occupations and leisure activities have a reduced risk of developing Alzheimer's disease (Scarmeas and Stern, 2003). In addition, animal research has shown that environmental enrichment enhances the plasticity and resilience of neurons and protects against neuronal dysfunction and degeneration in models of Alzheimer's disease (Lazarov et al., 2005). Similar to the beneficial effects of exercise on muscle cells, purposeful activity in neuronal circuits imposes a mild stress on the neurons. During such activity there is Na^+ and Ca^{2+} influx, increased production of free radicals, and increased energy (ATP and NAD^+) utilization. Normally, the neurons respond adaptively to the latter stresses via the activation of several different signaling pathways (Arumugam et al., 2006). For example, Ca^{2+} influx activates Ca^{2+}/calmodulin-dependent kinases and mitogen-activated protein kinases, resulting in the activation of transcription factors (e.g., CREB and NF-κB), which induce the expression of genes encoding Ca^{2+}-regulating proteins (glutamate receptors and Ca^{2+}-binding proteins) and cytoprotective neurotrophic factors such as BDNF (Balazs, 2006; Wang et al., 2007). Oxidative stress can activate the Nrf-2–ARE (antioxidant response element) pathway, resulting in the production of proteins that protect cells against oxidative stress, including heme oxygenase-1 (HO-1) and an oxidoreductase called NQO1 (Kang et al., 2005; Hyun et al., 2006). Cellular energy depletion activates AMPK, which, in turn, phosphorylates protein substrates that collectively function to increase energy availability while reducing energy demand (Viollet et al., 2007). Thus, activity in neuronal circuits enhances the ability of neurons to cope with Ca^{2+} influx and oxidative and metabolic stress.

Neurons in the couch potato's brain are relatively unchallenged, however, such that the hormetic pathways described in the preceding paragraph are largely untapped. Accordingly, it is likely that the neurons in the brain of a couch potato are caught with their guard down, and so are vulnerable to the aging process and associated diseases. Studies of animal models support the notion that neurons in the brains of individuals who live an unchallenged lifestyle are more likely to succumb in the settings of injury and disease. For example, exposure of rats to an enriched environment and positive social interactions resulted in an improved functional recovery following a stroke (Johansson and Ohlsson, 1996). Similarly, environmental enrichment attenuates cognitive deficits and reduces neuronal degeneration in a rat model of traumatic brain injury (Passineau et al., 2001). Of interest, it was recently reported that enhanced cognitive activity, beyond normal social activity or regular exercise, is required to protect Alzheimer's mice against cognitive impairment (Cracchiolo et al., 2007). Other studies of mouse models of Alzheimer's disease suggest that cognitive activity is as beneficial, or even more beneficial, in protecting neurons against dysfunction and damage (Wolf et al., 2006). The kinds of activities in which the typical couch potato engages are passive—watching television or sporting events. There is increasing evidence that these activities may not activate neuronal circuits in ways that induce adaptive stress response pathways. For example, one study revealed a positive correlation between hours of television watching in midlife and the risk of developing Alzheimer's disease in old age (Lindstrom et al., 2005). A study of twins in Sweden demonstrated that the risk of Alzheimer's disease was greater in the twin with a lower level of education (Gatz et al., 2001). Another epidemiological study concluded that individuals who engage in fewer activities in midlife are at increased risk of Alzheimer's disease (Friedland et al., 2001). It can be proposed that the disengagement phenotype of the couch potato is detrimental to the brain because it does not challenge/exercise neurons in amounts required to activate hormetic signaling pathways.

Implications of Hormesis for the Future of the Couch Potato

Of course the number of couch potatoes could be reduced if there were ways to motivate them to exercise, cut back on calorie intake, and engage in intellectually challenging activities. It takes effort to avoid becoming a couch potato and even more effort to transform from a couch potato to an active lifestyle. Beyond the adverse health consequences for the couch potato, his or her children are likely to emulate the lifestyle, and so couch potatoes proliferate. However, a major reason for the explosion of the couch potato lifestyle during the last half-century is related to the technological advances that have reduced the need for physical and mental activities and promoted the consumption of high-calorie fast foods. Manual labor has been replaced by machines on farms, in manufacturing, in construction, and even in home and yard care. Instead of washing dishes by hand or mowing the lawn with a push mower or raking leaves, we watch television or surf the Internet.

Scientific advances may, in the future, make it possible to accrue the benefits of exercise, dietary energy restriction, and an intellectually challenging lifestyle without actually engaging in these activities. It has been possible, in the laboratory, to activate hormetic pathways and so improve health using dietary supplements and drugs. The antidiabetic drug metformin induces mild stress on cells, resulting in the activation of AMPK, thereby enhancing glucose uptake into muscles and resulting in a reduction in blood glucose levels (Shaw et al., 2005). Dietary supplementation with 2-deoxy-D-glucose improved the resistance of the cardiovascular and neuroendocrine systems to stress in rats (Wan et al., 2004), and administration of 2-deoxy-D-glucose protected neurons in the brain against a stroke in rats (Yu and Mattson, 1999). The mechanism by which 2-deoxy-D-glucose protects cells is by imposing a mild metabolic stress that results in the induction of the expression of heat-shock proteins. Specific phytochemicals have also been shown to benefit health by inducing an adaptive stress response. Sulforaphane, which is present in broccoli, activates the Nrf-2–ARE pathway and thereby protects cells against cancer (Juge et al., 2007). Resveratrol, a chemical in red grapes and wine, has been shown to activate an enzyme called SIRT-1 that engages adaptive cellular stress responses that may protect cells in the cardiovascular and nervous systems (Baur and Sinclair, 2006).

Acknowledgments This work was supported by the Intramural Research Program of the National Institute on Aging, National Institutes of Health. The authors wish to thank K. C. Alexander for the preparation of Fig. 1

References

Ahmet I, Wan R, Mattson MP, Lakatta EG, Talan M (2005) Cardioprotection by intermittent fasting in rats. Circulation 112: 3115–3121.

Antuna-Puente B, Feve B, Fellahi S, Bastard JP (2008) Adipokines: the missing link between insulin resistance and obesity. Diabetes Metab 34: 2–11.

Arumugam TV, Gleichmann M, Tang SC, Mattson MP (2006) Hormesis/preconditioning mechanisms, the nervous system and aging. Ageing Res Rev 5: 165–178.

Astrup A, Dyerberg J, Selleck M, Stender S (2008) Nutrition transition and its relationship to the development of obesity and related chronic diseases. Obes Rev 9: S48–S52.

Baggio LL, Drucker DJ (2006) Therapeutic approaches to preserve islet mass in type 2 diabetes. Annu Rev Med 57: 265–281.

Balazs R (2006) Trophic effect of glutamate. Curr Top Med Chem 6: 961–968.

Bau AM, Ernert A, Schenk L, Wiegand S, Martus P, Grueters-Kieslich A, Krude H (2009) Is there a further acceleration in the age at onset of menarche? A cross-sectional study in 1840 school children focusing on age and bodyweight at the onset of menarche. Eur J Endocrinol 160: 107–113.

Baur JA, Sinclair DA (2006) Therapeutic potential of resveratrol: the in vivo evidence. Nat Rev Drug Discov 5: 493–506.

Benton CR, Wright DC, Bonen A (2008) PGC-1alpha-mediated regulation of gene expression and metabolism: implications for nutrition and exercise prescriptions. Appl Physiol Nutr Metab 33: 843–862.

Bose AK, Mocanu MM, Carr RD, Brand CL, Yellon DM (2005) Glucagon-like peptide 1 can directly protect the heart against ischemia/reperfusion injury. Diabetes 54: 146–151.

Calvo JA, Daniels TG, Wang X, Paul A, Lin J, Spiegelman BM, Stevenson SC, Rangwala SM (2008) Muscle-specific expression of PPARgamma coactivator-1alpha improves exercise performance and increases peak oxygen uptake. J Appl Physiol 104: 1304–1312.

CDC. Diabetes 2008: disabling disease to double by 2050. Atlanta, GA: U.S. Department of Health and Human Services, CDC; 2008. Available at: http://www.cdc.gov/nccdphp/publications/aag/pdf/diabetes.pdf.

Cracchiolo JR, Mori T, Nazian SJ, Tan J, Potter H, Arendash GW (2007) Enhanced cognitive activity—over and above social or physical activity—is required to protect Alzheimer's mice against cognitive impairment, reduce Abeta deposition, and increase synaptic immunoreactivity. Neurobiol Learn Mem 88: 277–294.

Diano S, Farr SA, Benoit SC, McNay EC, da Silva I, Horvath B, Gaskin FS, Nonaka N, Jaeger LB, Banks WA, Morley JE, Pinto S, Sherwin RS, Xu L, Yamada KA, Sleeman MW, Tschöp MH, Horvath TL (2006) Ghrelin controls hippocampal spine synapse density and memory performance. Nat Neurosci 9: 381–388.

Dixit VD, Taub DD (2005) Ghrelin and immunity: a young player in an old field. Exp Gerontol 40:900–910.

Doyle ME, Egan JM (2007) Mechanisms of action of glucagon-like peptide 1 in the pancreas. Pharmacol Ther 113: 546–593.

Duan W, Guo Z, Jiang H, Ware M, Li XJ, Mattson MP (2003) Dietary restriction normalizes glucose metabolism and BDNF levels, slows disease progression, and increases survival in huntingtin mutant mice. Proc Natl Acad Sci USA 100: 2911–2916.

Duan W, Mattson MP (1999) Dietary restriction and 2-deoxyglucose administration improve behavioral outcome and reduce degeneration of dopaminergic neurons in models of Parkinson's disease. J Neurosci Res 57: 195–206.

During MJ, Cao L, Zuzga DS, Francis JS, Fitzsimons HL, Jiao X, Bland RJ, Klugmann M, Banks WA, Drucker DJ, Haile CN (2003) Glucagon-like peptide-1 receptor is involved in learning and neuroprotection. Nat Med 9: 1173–1179.

Dyck DJ (2005) Leptin sensitivity in skeletal muscle is modulated by diet and exercise. Exerc Sport Sci Rev 33: 189–194.

Eggermont L, Swaab D, Luiten P, Scherder E. (2006) Exercise, cognition and Alzheimer's disease: more is not necessarily better. Neurosci Biobehav Rev 30: 562–575.

Everitt AV, Le Couteur DG (2007) Life extension by calorie restriction in humans. Ann N Y Acad Sci 1114: 428–433.

Fontana L, Klein S (2007) Aging, adiposity, and calorie restriction. JAMA 297: 986–994.

Friedland RP, Fritsch T, Smyth KA, Koss E, Lerner AJ, Chen CH, Petot GJ, Debanne SM (2001) Patients with Alzheimer's disease have reduced activities in midlife compared with healthy control-group members. Proc Natl Acad Sci USA 98: 3440–3445.

Fruhbeck G (2006) Intracellular signalling pathways activated by leptin. Biochem J 393: 7–20.

García EA, Korbonits M (2006) Ghrelin and cardiovascular health. Curr Opin Pharmacol 6: 142–147.

Gatz M, Svedberg P, Pedersen NL, Mortimer JA, Berg S, Johansson B (2001) Education and the risk of Alzheimer's disease: findings from the study of dementia in Swedish twins. J Gerontol B Psychol Sci Soc Sci 56: P292–P300.

Gomez-Pinilla F (2008) The influences of diet and exercise on mental health through hormesis. Ageing Res Rev 7: 49–62.

Greer EL, Brunet A (2005) FOXO transcription factors at the interface between longevity and tumor suppression. Oncogene 24: 7410–7425.

Halagappa VK, Guo Z, Pearson M, Matsuoka Y, Cutler RG, Laferla FM, Mattson MP (2007) Intermittent fasting and caloric restriction ameliorate age-related behavioral deficits in the triple-transgenic mouse model of Alzheimer's disease. Neurobiol Dis 26: 212–220.

Hamilton MT, Hamilton DG, Zderic TW (2007) Role of low energy expenditure and sitting in obesity, metabolic syndrome, type 2 diabetes, and cardiovascular disease. Diabetes 56: 2655–2667.

Handschin C, Chin S, Li P, Liu F, Maratos-Flier E, Lebrasseur NK, Yan Z, Spiegelman BM (2007) Skeletal muscle fiber-type switching, exercise intolerance, and myopathy in PGC-1alpha muscle-specific knock-out animals. J Biol Chem 282: 30014–30021.

Handschin C, Spiegelman BM (2008) The role of exercise and PGC1alpha in inflammation and chronic disease. Nature 454: 463–469.

Hoyda TD, Samson WK, Ferguson AV (2009) Adiponectin depolarizes parvocellular paraventricular nucleus neurons controlling neuroendocrine and autonomic function. Endocrinology 150: 832–840.

Hursting SD, Lavigne JA, Berrigan D, Perkins SN, Barrett JC (2003) Calorie restriction, aging, and cancer prevention: mechanisms of action and applicability to humans. Annu Rev Med 54: 131–152.

Hyun DH, Hernandez JO, Mattson MP, de Cabo R (2006) The plasma membrane redox system in aging. Ageing Res Rev 5: 209–220.

Johansson BB, Ohlsson AL (1996) Environment, social interaction, and physical activity as determinants of functional outcome after cerebral infarction in the rat. Exp Neurol 139: 322–327.

Johnson JB, Summer W, Cutler RG, Martin B, Hyun DH, Dixit VD, Pearson M, Nassar M, Telljohann R, Maudsley S, Carlson O, John S, Laub DR, Mattson MP (2007) Alternate day calorie restriction improves clinical findings and reduces markers of oxidative stress and inflammation in overweight adults with moderate asthma. Free Radic Biol Med 42: 665–674.

Juge N, Mithen RF, Traka M (2007) Molecular basis for chemoprevention by sulforaphane: a comprehensive review. Cell Mol Life Sci 64: 1105–1127.

Kang KW, Lee SJ, Kim SG (2005) Molecular mechanism of nrf2 activation by oxidative stress. Antioxid Redox Signal 7: 1664–1673.

Lajunen HR, Keski-Rahkonen A, Pulkkinen L, Rose RJ, Rissanen A, Kaprio J (2007) Are computer and cell phone use associated with body mass index and overweight? A population study among twin adolescents. BMC Public Health 26(7): 24.

Lazarov O, Robinson J, Tang YP, Hairston IS, Korade-Mirnics Z, Lee VM, Hersh LB, Sapolsky RM, Mirnics K, Sisodia SS (2005) Environmental enrichment reduces Abeta levels and amyloid deposition in transgenic mice. Cell 120: 701–713.

Li Y, Perry T, Kindy MS, Harvey BK, Tweedie D, Holloway HW, Powers K, Shen H, Egan JM, Sambamurti K, Brossi A, Lahiri DK, Mattson MP, Hoffer BJ, Wang Y, Greig NH (2009) GLP-1 receptor stimulation preserves primary cortical and dopminergic neurons in cellular and rodent models of stroke and Parkinsonism. Proc Natl Acad Sci USA 106: 1285–1290.

Lin J, Wu H, Tarr PT, Zhang CY, Wu Z, Boss O, Michael LF, Puigserver P, Isotani E, Olson EN, Lowell BB, Bassel-Duby R, Spiegelman BM (2002) Transcriptional co-activator PGC-1 alpha drives the formation of slow-twitch muscle fibres. Nature 418: 797–801.

Lindstrom HA, Fritsch T, Petot G, Smyth KA, Chen CH, Debanne SM, Lerner AJ, Friedland RP (2005) The relationships between television viewing in midlife and the development of Alzheimer's disease in a case–control study. Brain Cogn 58: 157–165.

Luchsinger JA, Tang MX, Shea S, Mayeux R (2002) Caloric intake and the risk of Alzheimer disease. Arch Neurol 59: 1258–1263.

Mager DE, Wan R, Brown M, Cheng A, Wareski P, Abernethy DR, Mattson MP (2006) Caloric restriction and intermittent fasting alter spectral measures of heart rate and blood pressure variability in rats. FASEB J 20: 631–637.

Martin B, Golden E, Carlson OD, Pistell P, Zhou J, Kim W, Frank BP, Thomas S, Chadwick WA, Greig NH, Bates GP, Sathasivam K, Bernier M, Maudsley S, Mattson MP, Egan JM (2009) Exendin-4 improves glycemic control, ameliorates brain and pancreatic pathologies and extends survival in a mouse model of Huntington's disease. Diabetes 58: 318–328.

Martin B, Pearson M, Kebejian L, Golden E, Keselman A, Bender M, Carlson O, Egan J, Ladenheim B, Cadet JL, Becker KG, Wood W, Duffy K, Vinayakumar P, Maudsley S, Mattson MP (2007) Sex-dependent metabolic, neuroendocrine, and cognitive responses to dietary energy restriction and excess. Endocrinology 148: 4318–4333.

Masoro EJ (2005) Overview of caloric restriction and ageing. Mech Ageing Dev 126: 913–922.

Maswood N, Young J, Tilmont E, Zhang Z, Gash DM, Gerhardt GA, Grondin R, Roth GS, Mattison J, Lane MA, Carson RE, Cohen RM, Mouton PR, Quigley C, Mattson MP, Ingram DK (2004) Caloric restriction increases neurotrophic factor levels and attenuates neurochemical and behavioral deficits in a primate model of Parkinson's disease. Proc Natl Acad Sci USA 101: 18171–18176.

Mattison JA, Roth GS, Lane MA, Ingram DK (2007) Dietary restriction in aging nonhuman primates. Interdiscip Top Gerontol 35: 137–158.

Mattson MP (2005) Energy intake, meal frequency, and health: a neurobiological perspective. Annu Rev Nutr 25: 237–260.

Mattson MP, Wan R (2005) Beneficial effects of intermittent fasting and caloric restriction on the cardiovascular and cerebrovascular systems. J Nutr Biochem 16: 129–137.

McAlpine DA, Manohar CU, McCrady SK, Hensrud D, Levine JA (2007) An office-place stepping device to promote workplace physical activity. Br J Sports Med 41: 903–907.

Nestle M, Jacobson MF (2000) Halting the obesity epidemic: a public health policy approach. Public Health Rep 115: 12–24.

Ouchi N, Shibata R, Walsh K (2006) Cardioprotection by adiponectin. Trends Cardiovasc Med 16: 141–146.

Passineau MJ, Green EJ, Dietrich WD (2001) Therapeutic effects of environmental enrichment on cognitive function and tissue integrity following severe traumatic brain injury in rats. Exp Neurol 168: 373–384.

Perry T, Holloway HW, Weerasuriya A, Mouton PR, Duffy K, Mattison JA, Greig NH (2007) Evidence of GLP-1-mediated neuroprotection in an animal model of pyridoxine-induced peripheral sensory neuropathy. Exp Neurol 203: 293–301.

Radak Z, Chung HY, Koltai E, Taylor AW, Goto S (2008) Exercise, oxidative stress and hormesis. Ageing Res Rev 7: 34–42.

Redman LM, Martin CK, Williamson DA, Ravussin E (2008) Effect of caloric restriction in non-obese humans on physiological, psychological and behavioral outcomes. Physiol Behav 94: 643–648.

Scarmeas N, Stern Y (2003) Cognitive reserve and lifestyle. J Clin Exp Neuropsychol 25: 625–633.

Shaw RJ, Lamia KA, Vasquez D, Koo SH, Bardeesy N, Depinho RA, Montminy M, Cantley LC (2005) The kinase LKB1 mediates glucose homeostasis in liver and therapeutic effects of metformin. Science 310: 1642–1646.

Shields M, Tremblay MS (2008) Sedentary behaviour and obesity. Health Rep 19: 19–30.

Shinmura K, Tamaki K, Saito K, Nakano Y, Tobe T, Bolli R (2007) Cardioprotective effects of short-term caloric restriction are mediated by adiponectin via activation of AMP-activated protein kinase. Circulation 116: 2809–2817.

Stern Y (2006) Cognitive reserve and Alzheimer disease. Alzheimer Dis Assoc Disord 20: 112–117.

Stranahan AM, Lee K, Becker KG, Zhang Y, Maudsley S, Martin B, Cutler RG, Mattson MP(2008) Hippocampal gene expression patterns underlying the enhancement of memory by running in aged mice. Neurobiol Aging 2008 Dec 11 [Epub ahead of print].

Sun Y, Asnicar M, Smith RG (2007) Central and peripheral roles of ghrelin on glucose homeostasis. Neuroendocrinology 86: 215–228.

Tilg H, Wolf AM (2005) Adiponectin: a key fat-derived molecule regulating inflammation. Expert Opin Ther Targets 9: 245–251.

Towler MC, Hardie DG (2007) AMP-activated protein kinase in metabolic control and insulin signaling. Circ Res 100: 328–341.

Vincent HK, Innes KE, Vincent KR (2007) Oxidative stress and potential interventions to reduce oxidative stress in overweight and obesity. Diabetes Obes Metab 9: 813–839.

Viollet B, Mounier R, Leclerc J, Yazigi A, Foretz M, Andreelli F (2007) Targeting AMP-activated protein kinase as a novel therapeutic approach for the treatment of metabolic disorders. Diabetes Metab 33: 395–402.

Vu V, Riddell MC, Sweeney G (2007) Circulating adiponectin and adiponectin receptor expression in skeletal muscle: effects of exercise. Diabetes Metab Res Rev 23: 600–611.

Wan R, Camandola S, Mattson MP (2004) Dietary supplementation with 2-deoxy-D-glucose improves cardiovascular and neuroendocrine stress adaptation in rats. Am J Physiol Heart Circ Physiol 287: H1186–H1193.

Wang JQ, Fibuch EE, Mao L (2007) Regulation of mitogen-activated protein kinases by glutamate receptors. J Neurochem 100: 1–11.

Wolf SA, Kronenberg G, Lehmann K, Blankenship A, Overall R, Staufenbiel M, Kempermann G (2006) Cognitive and physical activity differently modulate disease progression in the amyloid precursor protein (APP)-23 model of Alzheimer's disease. Biol Psychiatry 60: 1314–1323.

Yu ZF, Mattson MP (1999) Dietary restriction and 2-deoxyglucose administration reduce focal ischemic brain damage and improve behavioral outcome: evidence for a preconditioning mechanism. J Neurosci Res 57: 830–839.

Hormesis and Aging

Suresh I.S. Rattan and Dino Demirovic

Abstract Mild stress-induced hormetic stimulation of protective mechanisms in cells and organisms can result in potential antiaging effects. Detailed molecular mechanisms that bring about the hormetic effects are being increasingly understood and comprise a cascade of stress response and other pathways of maintenance and repair. Although the extent of immediate hormetic effects after exposure to a particular stress may only be moderate, the chain of events following initial hormesis leads to biologically amplified effects that are much larger, synergistic, and pleiotropic. A consequence of hormetic amplification is an increase in the homeodynamic space of a living system in terms of increased defense capacity and reduced load of damaged macromolecules. Hormetic strengthening of the homeodynamic space provides wider margins for metabolic fluctuation, stress tolerance, adaptation, and survival. Hormesis thus counterbalances the progressive shrinkage of the homeodynamic space that is the ultimate cause of aging, diseases, and death. Healthy aging may be achieved by hormesis through mild and periodic but not severe or chronic physical and mental challenges and by the use of nutritional hormesis incorporating mild stress-inducing molecules called hormetins.

Keywords Antiaging · Homeostasis · Longevity · Skin · Stress

Introduction

Because the harmful effects of severe and chronic stress have long overshadowed the beneficial hormetic effects of low-level stress, the application of hormesis in aging research and therapy is a relatively recent development. The paradigm for considering the applicability of hormesis in aging intervention is the well-documented beneficial effect of moderate exercise, which at a biochemical level results in the

---

S.I.S. Rattan (✉)
Laboratory of Cellular Aging, Department of Molecular Biology, University of Aarhus, DK 8000 Aarhus C, Denmark
e-mail: rattan@mb.au.dk

M.P. Mattson, E.J. Calabrese, *Hormesis*, DOI 10.1007/978-1-60761-495-1_9,

production of potentially harmful substances such as free radicals, acids, and aldehydes. Thus, it was hypothesized that if aging systems are deliberately exposed to mild stress, this should lead to achieving beneficial hormetic effects, including health- and longevity-promoting effects. Hormesis in aging is therefore defined as the life-supporting beneficial effects resulting from the cellular responses to single or multiple rounds of mild stress (Rattan, 2001a, b, 2004, 2005).

Here we review and analyze the published literature on various physical, chemical, and biological conditions that are known to be potentially harmful at high doses but that at lower doses have the effects of slowing down aging and/or prolonging the lifespan of cells and organisms. Table 1 lists the main stresses that have been shown to have aging- and longevity-modulatory effects in various systems.

Table 1 Various Types of Stresses Tested for Their Antiaging Effects

Physical stress
Thermal
Hypergravity
Radiation
Exercise
Biological stress
Dietary restriction
Dietary components
Natural hormetins
Chemical stress
Minerals
Heavy metals
Pro-oxidants
Synthetic hormetins

However, it is important to point out that so far only a few studies have been performed with the specific aim of testing the applicability of hormesis in aging, for example, those using thermal stress and hypergravity as hormetic agents. For most other studies that have been interpreted to involve hormesis as the mode of action of the stressful conditions used in those experiments, these conclusions are generally derived in retrospective analyses. Such studies include the effects of radiation, exercise, pro-oxidants, nutritional components, and food restriction. However, to fully appreciate the rationale for using hormesis as a modulator of aging and longevity, we first provide a brief overview of the biological understanding of aging, which is considered to be no longer an unresolved problem in biology (Hayflick, 2007; Holliday, 2006).

Recapitulating the Biological Basis of Aging

Biogerontology—the study of the biological basis of aging—has unveiled mysteries of aging by describing age-related changes in organisms, organs, tissues, cells, and macromolecules. The large body of descriptive data has led two of the pioneers of modern biogerontology, Leonard Hayflick and Robin Holliday, to declare that aging is no longer an unsolved problem in biology (Hayflick, 2007; Holliday, 2006). This declaration does not mean that there are no remaining descriptive data on aging and that every piece of information about aging in every biological system has been gathered. The bold assertion by Hayflick and Holliday underlines the fact that the biological basis of aging is well understood and a distinctive framework has been established that will not be altered significantly with additional descriptive data. Based on the large body of descriptive data, certain general principles of aging and longevity can be clearly formulated, and these can be the basis for translational research and interventions toward achieving a healthy old age (Table 2).

Table 2 General Principles of Aging and Longevity Derived from Modern Biogerontological Research

- *Life history principle:* Aging is an emergent phenomenon seen primarily in protected environments that allows survival beyond the natural lifespan of a species, termed "essential lifespan" (ELS) (Rattan, 2000a, b; Rattan and Clark, 2005).
- *Differential principle:* The progression and rate of aging is different in different species, in organisms within a species, in organs and tissues within an organism, in cell types within a tissue, in subcellular compartments within a cell type, and in macromolecules within a cell.
- *Mechanistic principle*: Aging is characterized by a progressive accumulation of molecular damage in nucleic acids, proteins, and lipids. The inefficiency and failure of maintenance, repair, and turnover pathways is the main cause of age-related accumulation of damage.
- *Nongenetic principle:* There is no fixed and rigid genetic program that determines the exact duration of survival of an organism, and there are no "gerontogenes" whose sole function is to cause aging and to determine precisely the lifespan of an organism.

Thus, aging has many facets, and almost all the experimental data suggest that aging is an emergent, epigenetic, meta-phenomenon that is not controlled by a single mechanism. Although individually no tissue, organ, or system becomes functionally exhausted even in very old organisms, it is their combined interaction and interdependence that determines the survival of the whole. A combination of genes, milieu, and chance determines the course and consequences of aging and the duration of survival of an individual (Rattan, 2007b).

There is much supporting evidence for the theory that the survival and longevity of a species are a function of the ability of its maintenance and repair mechanisms to keep up with damage and wear and tear. All living systems have the intrinsic ability to respond to, counteract, and adapt to external and internal sources of disturbance. The traditional conceptual model to describe this property is homeostasis, which has dominated biology, physiology, and medicine since the 1930s. However, advances in our understanding of the processes of biological growth, development, maturation, reproduction, and aging, senescence, and death have led to the realization that the homeostasis model as an explanation is seriously incomplete. The main reason for the incompleteness of the homeostasis model is its defining principle of "stability through constancy," which does not take into account the new themes, such as cybernetics, control theory, catastrophe theory, chaos theory, information, and interaction networks, that comprise and underlie the modern biology of complexity

(Rattan, 2007a). Since the 1990s, the term homeodynamics has been increasingly used to account for the fact that the internal milieu of complex biological systems is not permanently fixed, is not at equilibrium, and is a dynamic regulation and interaction among various levels of organization (Yates, 1994).

Aging, senescence, and death are the final manifestations of unsuccessful homeostasis or failure of homeodynamics (Holliday, 2007; Rattan, 2006). A wide range of molecular, cellular, and physiological pathways of repair are well known, and these range from multiple pathways of nuclear and mitochondrial DNA repair to free radical counteracting mechanisms, protein turnover and repair, detoxification mechanisms, and other processes, including immune and stress responses. All of these processes involve numerous genes whose products and interactions give rise to the "homeodynamic space" or "buffering capacity that is the ultimate determinant of an individual's chance and ability to survive and maintain a healthy state (Holliday, 2007; Rattan, 2006). A progressive shrinking of the homeodynamic space is the hallmark of aging and the cause of age-related diseases.

Figure 1 is a pictorial representation of the concept of homeodynamic space and the consequences of its shrinkage during aging. In a normal, healthy, young individual, the complex network of maintenance and repair systems (MRS) constitutes a functional homeodynamic space. Because no MRS can be 100% efficient 100% of the time, even in a young system, there is a probability of incomplete homeodynamics, giving rise to a zone of vulnerability, manifested in age-independent diseases and mortality. However, a progressive accumulation of molecular damage and its effects on the interacting molecular networks leads to the reduction in the functional homeodynamic space and effectively increases the vulnerability zone, thus allowing for the occurrence and emergence of age-related diseases. Alzheimer's disease, cancer, cataract, diabetes type 2, osteoporosis, Parkinson's disease, sarcopenia, and other age-related diseases are the result of an individual's reduced homeodynamic space.

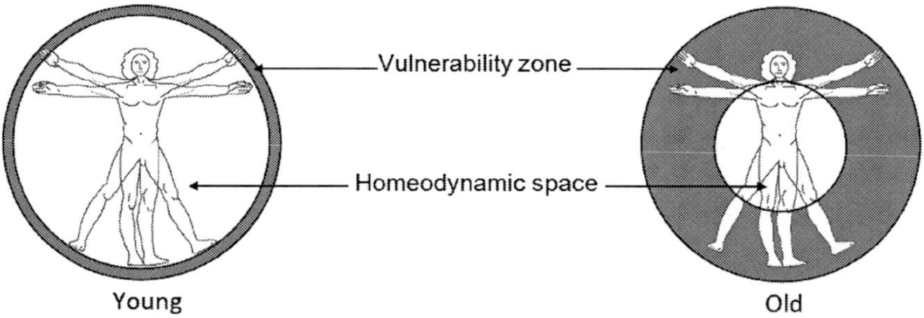

Fig. 1 Pictorial representation of the concept of homeodynamic space, whose progressive shrinkage due to the accumulation of molecular damage leads to an increase in the area of vulnerability zone in the elderly, and hence to the occurrence and emergence of age-related diseases

A critical component of the homeodynamic property of living systems is their capacity to respond to stress. In this context, the term "stress" is defined as a signal generated by any physical, chemical, or biological factor (stressor) that in a living system initiates a series of events to enable the organism to counteract, adapt, and survive. Although a successful and over-compensatory response to low doses of stressors improves the overall homeodynamics of cells and organisms, an incomplete or failed homeodynamic response leads to the damaging and harmful effects of stress, including death. It is this homeodynamic space as a whole or the individual components of the homeodynamic machinery that are the targets of hormetic interventions.

Thermal Hormesis in Aging

Thermal Hormesis in Organisms

Temperature stress, especially high-temperature–induced heat shock (HS), has been widely used with the specific aim of testing and applying hormesis in aging research and interventions. One of the main reasons for choosing HS as a hormetic agent is that HS acts through an evolutionarily highly conserved stress response pathway, known as the heat-shock response, the molecular basis of which is well understood (Sun and MacRae, 2005; Verbeke et al., 2001b). Effects of mild and severe HS have been tested on yeast, nematodes, fruit flies, and rodent and human cells. For example, wild-type and *age-1* long-lived mutant hermaphrodite *Caenorhabditis elegans* exposed for 3 to 24 hours to 30°C exhibited a significant increase in mean lifespan compared to controls (Johnson, 2002; Lithgow et al., 1995). Similarly, a 6-hour exposure at 30°C of wild-type worms increased their lifespan, but no effect was found after exposures of 2 or 4 hours (Yokoyama et al., 2002). Furthermore, studies of *C. elegans* subjected to 35°C HS for different durations showed that HS not longer than 2 hours resulted in an extension of lifespan (Butov et al., 2001; Michalski et al., 2001; Yashin et al., 2001). In a study of multiple stresses in *C. elegans* an extension of lifespan after 1 and 2 hours of HS at 35°C was reported (Cypser and Johnson, 2002, 2003). In another study performed on *C. elegans* it was observed that repeated mild HS throughout life had a larger effect on lifespan than a single mild HS early in life, and the effect was related to the levels of heat-shock protein (HSP) expression (Olsen et al., 2006).

In the case of fruit flies, virgin males of inbred lines of *Drosophila melanogaster* exhibited an increase in mean lifespan and lower mortality rates during several weeks after a heat treatment of 36°C for 70 min (Khazaeli et al., 1997). It has also been shown that wild-type *D. melanogaster* exposed to 37°C for 5 minutes a day, 5 days a week for 1 week lived on average 2 days longer than the control flies (Le Bourg et al., 2001). Longer exposures had either no effect or a negative effect on lifespan. In another study on *D. melanogaster*, the exposure of young flies to four rounds of mild HS at 34°C significantly increased the average and maximum lifespan of female flies and increased their resistance to potentially lethal HS (Hercus et al., 2003). Of interest, the beneficial effects of HS in *Drosophila* did not entirely depend on the continuous presence of HSP but were observed long after newly synthesized HSP had disappeared, indicating the involvement of a cascade of poststress events in hormesis (Sørensen et al., 2008). Furthermore, the hormetic effects of HS appear to occur to different extents in male and female *Drosophila*, which may be due to the fact that females have to trade off stress resistance and reproduction (Sørensen et al., 2008).

Studies have also been performed on the effect of subjecting transgenic *D. melanogaster* overexpressing the inducible HSP70 to 20 minutes at 36°C (Minois et al., 2001; Minois and Vaynberg, 2002). In the control "parental" line, such exposure significantly increased the lifespan of both virgin flies kept in groups and mated flies. In individually kept flies, the same trend was observed but was statistically not significant. No beneficial effect of such HS has been seen in the transgenic lines, which may be suggestive of upper limits of modulating HS responses (Minois and Vaynberg, 2002). In addition to the high temperature, there is some evidence demonstrating that cold shocks at young age increased the longevity and survival of *Drosophil*a at high temperature and increased longevity of *Drosophila* after cold stress–induced hardening (Le Bourg, 2008; Overgaard et al., 2005). In the case of mammals, irradiated and nonirradiated mice that were given intermittent cold shocks showed lower rates of mortality in the irradiated mice. Longer lifespans were observed in thermally stressed nonirradiated males and irradiated females (Minois, 2000). Similarly, rats kept in water set at 23°C, 4 hours a day, 5 days a week, had a 5% increase in average lifespan and diminished occurrence of age-related diseases (Holloszy and Smith, 1986).

Thermal Hormesis in Human Cells Undergoing Aging in Vitro

A series of studies performed in our labs tested the hormesis hypothesis of the beneficial effects of mild HS, using the Hayflick system of cellular aging of normal human cells in culture. Employing a mild stress regimen of exposing serially passaged human skin fibroblasts to 41°C for 1 hour twice a week throughout their replicative lifespan, we found several antiaging effects, which are listed in Table 3.

The choice of the repeated mild heat shock (RMHS) regimen was based on several pilot experiments performed for selecting conditions in which 30% of the maximal HS response was elicited without affecting cell growth and survival (Kraft et al., 2006; Rattan, 1998). This does not imply that these are the ideal hormetic conditions for these cells. Other combinations of dose and duration may well have similar or even better effects, but that issue remains to be investigated. Furthermore, we also showed that repeated mild HS at 41°C, but not the relatively severe HS at 42°C, increased the replicative lifespan and elevated and maintained the basal levels of MAP kinases JNK1, JNK2, and p38 in human skin fibroblasts (Nielsen et al., 2006).

To confirm the wider applicability of mild HS-induced hormesis in other human cell types, we also performed studies on normal human epidermal keratinocytes (NHEKs) and obtained results that were very similar to those for dermal fibroblasts. NHEK also showed a variety of cellular and biochemical hormetic antiaging effects on repeated exposure to mild HS at 41°C. These effects included maintenance of youthful cellular morphology, enhanced replicative lifespan, enhanced proteasomal

Table 3 A Summary of Results of Studies on the Antiaging Hormetic Effects of Repeated Mild Heat Shock on Human Skin Fibroblasts in Vitro

Characteristic	Hormetic Effect	Reference
Cellular phenotype		
Cell size	Reduced enlargement	(Rattan, 1998)
Cell morphology	Reduced irregularization	(Rattan, 1998)
Proliferative lifespan	20% increase	(Nielsen et al., 2006)
Wound healing	30% increase	(Rattan et al., 2009)
Cell physiological phenotype		
H_2O_2 decomposing ability	50%–140% increase	(Fonager et al., 2002)
Survival after H_2O_2 exposure	10%–18% increase	(Fonager et al., 2002)
Survival after ethanol exposure	10%–40% increase	(Fonager et al., 2002)
Survival after ultraviolet A exposure	5%–17% increase	(Fonager et al., 2002)
Molecular damage		
Glucation, furasine level	50%–80% reduction	(Verbeke et al., 2001a)
Glycoxidation level	10%–30% reduction	(Verbeke et al., 2001a)
Carboxymethyl-lysine–rich protein level	20%–85% reduction	(Verbeke et al., 2001a)
Lipofuscin pigment level	6%–29% reduction	(Verbeke et al., 2001a)
Protein carbonyl level	5%–40% reduction	(Verbeke et al., 2001a)
Reduced glutathione level	3-fold increase	(Verbeke et al., 2001a)
Oxidized glutathione level	2-fold reduction	(Verbeke et al., 2001a)
Induction of sugar-induced protein damage	10-fold reduction	(Verbeke et al., 2002)
Molecular mechanisms		
HSP27 level	20%–40% increase	(Fonager et al., 2002)
HSC70 level	20% increase	(Fonager et al., 2002)
HSP70 level	7- to 20-fold increase	(Fonager et al., 2002)
HSP90 level	50%–80% reduction	(Fonager et al., 2002)
Proteasome activities	40%–90% increase	(Beedholm et al., 2004)
11S activator content	2-fold increase	(Beedholm et al., 2004)
11S activator binding	2-fold increase	(Beedholm et al., 2004)
JNK1, JNK2 and p38 level	45%–70% increase	(Nielsen et al., 2006)

activity, and increased levels of HSPs (Rattan and Ali, 2007). In addition, we also studied the effects of HS on Na,K-ATPase and the sodium pump, whose content and activity were increased significantly after mild HS (Rattan and Ali, 2007). However, the molecular mechanisms and interactions that bring about the mild HS-induced increase in the amounts and activity of Na,K-ATPase and their consequences on other biochemical pathways during aging are yet to be elucidated. Notably, comparable hormetic effects could not be seen in NHEK repeatedly exposed to 43°C, which underlines the differences between the beneficial effects of mild stress and the harmful effects of severe stress.

We also observed that mild HS enhances the ability of serially passaged keratinocytes to enter into differentiation in the presence of calcium, as measured by the levels of differentiation markers involucrin, p38, and HSP27 (Berge et al., 2008). Another cell type in which we tested whether differentiation can be improved hormetically by RMHS is the telomerase-immortalized bone marrow mesenchymal stem cell-line hTERT-MSC. Single or multiple exposures to mild HS significantly enhanced the vitamin D–induced differentiation of hTERT-MSC into osteoblasts, as measured by determining the levels of osteoblastic markers alkaline phosphatase and mineralized matrix (Nørgaard et al., 2006). Although the mechanistic aspects of the hormetic effects of HS on the differentiation of human cells are yet to be elucidated, such studies pave the way for developing novel means for the maintenance and improvement of differentiation abilities of various cell types and thus preventing age-related alterations that lead to impairments such as thinning and excessive wrinkling of the skin and loss of bone mass leading to osteoporosis.

Other hormetic effects of mild HS on human cells that we have observed are improved wound healing and enhanced angiogenesis in vitro (Rattan et al., 2009). For example, HS-conditioned medium collected from one set of cultures after 6 hours post-HS at 41°C for 1 hour enhanced wound healing by 17% to 38% in a separate set of cells. This increase in wound healing by HS-conditioned medium was accompanied by a 68% increase in mobility and migration of cells and by about 54% enhanced elongation of individual cells. These studies indicate that mild HS induces the synthesis of one or more gene products secreted by the cells in the culture medium. Furthermore, these molecules can stimulate wound healing either as direct stimulants or as inhibitors of the negative modulators of wound healing, such as the plasminogen activator inhibitor PAI-1 (Kortlever and Bernards, 2006). However, the full range of secreted proteins, including HSPs, that may be responsible for enhanced wound healing and other biological effects are yet to be identified. We are analyzing various other molecular markers of cell migration, such as paxillin, talin, and focal adhesions, to elucidate the mechanisms of mild HS-induced improvements in wound healing.

Improved angiogenesis in vitro is another hormetic effect of mild HS that we have observed. Preexposure of normal human umbilical vein endothelial cells (HUVECs) to 1 hour of HS at 41°C or 42.5°C, followed by different periods of recovery at 37°C, had hormetic effects with respect to angiogenesis. Of interest, whereas the general extent and quality of the tubes formed by cells preexposed to 41°C were better than those in the controls, a preexposure at 42.5°C resulted in a relative worsening of tube structures, indicating that only mild stress has hormetic effects (Rattan et al., 2009). We are now attempting to elucidate whether the extent of hormetic effects of mild HS on angiogenesis are related to the levels of various HSPs synthesized during this period and what other pathways are involved in this. For example, there is some evidence that HSP90 stimulates tube formation by HUVEC via its role in enhancing the expression of the nitric oxide synthase (NOS) gene and the production of nitric oxide (Sun and Liao, 2004).

The molecular mechanisms through which the hormetic effects of mild HS are achieved remain to be elucidated. Although the general mechanisms of severe HS response are well understood, it is not clear whether there are any significant differences between mild HS, which has hormetic effects, and severe HS, which has deleterious effects (Park et al., 2005). It is likely that the physiological cost of stress in terms of energy utilization, molecular damage overload, and metabolic shift determines the difference between the outcomes of mild and severe stress. In addition, it is yet to be understood how the transient appearance of HSPs leads to biologically amplified hormetic effects at various other levels of cellular functioning, such as improved proteasome activity, enhanced resistance to other stresses, and maintenance of the cytoskeletal integrity.

Hypergravity Hormesis in Aging

Antiaging and life-prolonging hormetic effects of hypergravity have been studied in *Drosophila.* Whereas lifelong exposure to hypergravity decreases the lifespan in rodents and fruit flies, a 2-week exposure to 3 or 5 g at earlier stages in life resulted in an increase of 15% in the lifespan of male but not of female *D. melanogaster* (Minois, 2006). In addition to longevity, other physiological and behavioral parameters, such as fecundity, fertility, locomotor activity, antioxidant enzyme activity, HSP levels, and heat resistance, have also been analyzed. Except for increased survival of hypergravity-exposed *Drosophila* under heat stress, no other clear-cut patterns have been observed that can be associated with antiaging effects of transient hypergravity (Le Bourg, 2008). It is also not clear why the longevity-extending hormetic effects of hypergravity are restricted to male flies. Studies on checking the antiaging effects of hypergravity on any of the molecular biomarkers of aging, such as the level of macromolecular damage and other maintenance and repair pathways, have yet to be performed.

Radiation Hormesis in Aging

Radiation Hormesis in Insects

One of the earliest studies to show the life-prolonging effects of irradiation were performed on the flour beetle, *Tribolium confusum,* in which repeated exposure of beetles to low-dose radiation (LDR) of X-rays reduced their death rates as compared with those of unexposed organisms. Similar observations on the life-extending effects of γ-rays and X-rays on flour beetles were later reported by others (for references, see Rattan, 2008). Other insects used for similar studies are the housefly, *Musca domestica,* and the fruitfly, *Drosophila.* For example, whereas high doses of radiation decreased the lifespan, LDR extended the lifespan of fruit flies and of houseflies (Rattan, 2008). It has been argued that irradiation leads to female sterility and that the lifespan increase was an outcome of decreased fecundity. It was also shown that mutant females without ovaries did not exhibit increased lifespan after irradiation. The long-term consequences of the X-irradiation of *Drosophila* eggs demonstrated longevity hormesis in male flies exposed to 0.5 and 0.75 Gy, which also had smaller amounts of DNA segments resulting from cleavage in S1 nuclease–sensitive sites (Vaiserman et al., 2003; 2004a; Vaiserman et al., 2004b). One explanation given for the life-extending effects of LDR in insects is that irradiation induces stable epigenetic DNA modifications and enhanced DNA repair capacity (Vaiserman, 2008).

Radiation Hormesis in Rodents and Other Animals

Several studies have reported the hormetic effects of γ-rays on longevity in rats, mice, and guinea pigs (Calabrese and Baldwin, 2000; Caratero et al., 1998). Suppression of thymic lymphoma induction and prolongation of lifespan associated with immunological modification by chronic LDR in C57BL/6 mice have been reported (Ina and Sakai, 2005; Ina et al., 2005). In contrast, some earlier studies had failed to show an increase of lifespan after LDR. For instance, deuteron-irradiated mice exhibit higher mortality rates and lower lifespan in both sexes than nonirradiated ones (Ordy et al., 1967).

There are some data available on the effects of LDR on the survival and longevity of the nematode *C. elegans*. For example, an increase in the survival of *C. elegans* was sometimes observed after intermediate levels of irradiation (Johnson and Hartman, 1988). However, pretreatment with ultraviolet or ionizing radiation did not promote subsequent resistance or increased longevity of the worms exposed to other hormetic stresses, such as heat, hyperbaric oxygen, and pro-oxidants (Cypser and Johnson, 2002).

Radiation Hormesis in Humans

The adaptive response of human embryonic cells to low-dose γ-radiation has been shown to increase the replicative lifespan significantly (Watanabe et al., 1992). Similarly, human embryonic lung diploid fibroblasts sequentially irradiated with 1 Gy γ-rays had their replicative lifespan increased to some extent (Holliday, 1991). Hormetic effects of low-dose X-irradiation on the proliferative ability, genomic stability, and activation of mitogen-activated protein kinase pathways have been reported for other human diploid cells (Suzuki et al., 1998a, 2001; Suzuki et al., 1998b; Tsutsui et al., 1997).

In the case of humans, there are some claims that exposure to LDR has antiaging and other health benefits such as cancer prevention, but the demographic data are insufficient and inconclusive (Parsons, 2003; Wyngaarden and Pauwels, 1995). For example, although better survival and other beneficial effects of low to intermediate doses of atomic bomb radiation on Hiroshima and Nagasaki survivors have been claimed (Hayakawa et al., 1989; Mine et al., 1990; Okajima et al., 1985), these results were challenged by later analyses (Cologne and Preston, 2000). On the other hand, mortality rates of all workers in the U.K. Atomic Energy Authority were found to be lower than national rates (Atkinson et al., 2004). All-cause mortality and all-cause cancers (leukemia and prostate cancer) were also significantly lower for nuclear workers than for nonradiation workers (Atkinson et al., 2004). In an earlier study analyzing the varying levels of environmental radiation and incidence of cancer in different parts of India, it was reported that the cancer risk was invariably less in regions such as Kerala where the background radiation level is higher due to rich coastal thorium-monazite deposits (Nambi and Soman, 1987). Recent analyses of a series of 15-country international cohort studies of nuclear workers and of people living near nuclear reactors also indicates that radiation effects are not linear in terms of survival and incidence of cancer and other diseases, and that these effects may be further accentuated as a function of age, health status, and lifestyle variations (Cardis et al., 2007; Vrijheid et al., 2007).

Application of low-dose (1.2–1.8 Gy) total body irradiation (TBI) in the treatment of cancers, such as non-Hodgkin's lymphoma, is considered to be an example of radiation hormesis (Safwat, 2008). This is due to the fact that low-dose TBI did not kill cancer cells directly, but enhanced their removal by the immune system by increasing the proportion of cytotoxic T-lymphocytes, helper-inducer T-lymphocytes, and helper T-lymphocytes while decreasing the proportion of the suppressor-inducer T-lymphocytes and suppressor T-lymphocytes (Safwat, 2008). Although the exact mechanisms of how LDR brings about beneficial and longevity-promoting effects are not fully understood, there is evidence that LDR stimulates various repair and maintenance pathways as the cellular response to counteract the damage induced by radiation. These pathways include enhanced DNA repair, induction of DNA methylation, increased levels of antioxidative enzymes, and increased removal of damaged macromolecules (Rattan, 2008). It will be useful to design and perform studies aimed specifically to test the antiaging, anticancer, and longevity-promoting hormetic effects of stress-inducing levels of irradiation.

Calorie Restriction and Hormesis

Calorie restriction (CR) is the most commonly used intervention that has shown to extend the lifespan and slow down the onset of a wide range of age-related changes in a variety of organisms, including yeast, insects, rats, mice, and monkeys. Beneficial effects of other CR regimens, such as 25% and 8.5% chronic CR (Gomez et al., 2007) and intermittent CR (once or twice a week), have also been reported in animal studies (Anson et al., 2003; Martin et al., 2006). In the case of humans, some beneficial and health-promoting effects of CR have been reported. For example, long-term CR is reported to be highly effective in reducing the risk for atherosclerosis in humans (Fontana et al., 2004) and ameliorates the decline in diastolic function in humans (Meyer et al., 2006). An unintentional CR imposed on the participants of the Biosphere 2 experiment in 1991 also gave some indication of the beneficial effects of CR, as measured by several physiological, hematological, hormonal, and biochemical parameters (Walford et al., 2002). Similarly, unintentional chronic undernutrition and low body mass index have been shown to improve certain DNA repair parameters in peripheral blood lymphocytes of human subjects (Raji et al., 1998).

A relatively short-duration CR for 6 months has also been shown to have beneficial effects in humans by reducing fasting insulin levels, body temperature, and DNA damage (Heilbronn et al., 2006). Intermittent CR by periodic fasting has been shown to have a range of beneficial effects in rodents (Anson et al., 2003; Sharma and Kaur, 2005; Sogawa and Kubo, 2000). These observations make periodic CR more easily applicable and acceptable to humans, with several potential benefits, especially with respect to the prevention of neurodegenerative diseases with age (Arumugum et al., 2006; Martin et al., 2006). We have reported antiaging and lifespan-extending effects in serially passaged human skin fibroblasts by periodic partial (80%) fasting (once a week for 24 hours) by serum reduction, resulting in enhanced autophagy (Moore, 2008; Rattan, 2007b).

Hormesis has been suggested as a major explanation for the antiaging effects of CR by considering CR as a low-intensity stressor (Masoro, 2007). The evidence in support of the view that CR is a low-intensity stressor is its association with the increase in plasma levels of glucocorticoid steroid stress hormones (reviewed in Masoro, 2007). Another requirement for the hormesis hypothesis to explain the effects of CR is that CR should work through one or more pathways involved in stress response, molecular damage prevention and turnover, and metabolic regulation. There is significant evidence for the hormetic action of CR through the promotion of maintenance and repair pathways, which include increase in nucleotide excision repair, increase in the level of chaperones, increase in the level of proteasomal activities, enhancement of lysosomal autophagy, reduction in mitochondrial free radical generation and increase in mitochondrial uncoupling, and a shift in the metabolic regulation involving sirtuins and insulin-dependent pathways (Bonelli et al., 2008; Masoro, 2007; Rattan, 2008).

Exercise Hormesis

Some of the main molecular pathways involved in bringing about the adaptive and hormetic effects of exercise are activation of the nuclear factor NF-κB signaling cascade involving various stress kinases and antioxidant genes (Ji, 2008), enhanced anti-inflammatory responses, enhanced DNA repair, and increased degradation of damaged proteins and other macromolecules by proteasomal and lysosomal pathways (Radak et al., 2005; Short et al., 2004). Another pathway that is active in realizing the hormetic effects of exercise is the stress response or HSP synthesis pathway, in which the induction of various HSPs during and after exercise has a variety of beneficial biological effects (Atalay et al., 2004; Lancaster et al., 2004). Increased levels of HSPs provide several benefits, including protection against molecular damage occurrence and accumulation, which is a crucial aspect of aging (Radak et al., 2008a, b).

Nutritional Hormesis and Hormetins

Several dietary components, such as vitamins, antioxidants, trace elements, minerals, and ethanol, have been shown to have typical hormetic dose response (Calabrese, 2004; Mattson, 2008). All such compounds (natural or synthetic) that bring about biologically beneficial effects by acting through one or more pathways of maintenance and repair and of stress response are termed *hormetins* (Ali and Rattan, 2006; Rattan, 2008). The hormetic effects of various vitamins and macro- and microminerals, including iron, iodine, fluorine, selenium, and copper, have been reviewed (Hayes, 2007). In addition, the effects of zinc also show a typical hormetic dose response, and its beneficial effects are achieved through stress response–induced alterations in gene expression in various maintenance and repair pathways (Mocchegiani et al., 2006).

Dietary intake of moderate amounts of ethanol has been shown to have memory-enhancing beneficial effects in mice (Ritzmann et al., 1994). In the case of humans, consumption of moderate amounts of alcohol, combined with other positive lifestyle factors, has been associated with fourfold reduction in mortality (Khaw et al., 2008). The cardioprotective, antioxidative, and other beneficial effects of wine are considered to be due to flavonoid and nonflavonoid components, such as resveratrol (Corder et al., 2006), which also have a hormetic dose response. Resveratrol is considered to be a product of sunlight- and microbial-stress–induced hormetic response (Lamming et al., 2004). Several studies have reported the antiaging and longevity-enhancing effects of resveratrol in nematodes, *Drosophila*, and mice (Baur et al., 2006; Rogina and Helfand, 2004; Valenzano et al., 2006; Wood et al., 2004). Because reseveratrol's mode of action involves regulating various pathways of maintenance, repair, and induction of HSP synthesis, it is another example of a hormetin (Putics et al., 2008).

Other compounds that qualify to be called hormetins are various antioxidants, including components of spices and other medicinal plants. Almost all antioxidants show hormetic dose response and become pro-oxidants above certain doses. Furthermore, in some cases, such as α-lipoic acid and coenzyme Q10, it is their pro-oxidant activity in producing hydrogen peroxide that induces defensive responses (Linnane and Eastwood, 2006). Certain mimetics of superoxide dismutase claimed to have antiaging effects also appear to work through hormetic pathways by inducing oxidative stress response (Keany et al., 2004; Liu et al., 2003; Melov et al., 2000). Even DNA damage products, for example, thymidine dimers, have cytoprotective effects in the skin by inducing DNA repair pathways (Eller et al., 1997; Goukassian et al., 2004). Another secondary DNA damage product, N^6-furfuryladenine or kinetin, which is known to have antiaging effects in human cells and is widely used as a component of several skin care cosmetic products, may also work as a hormetin through stress-induced hormetic pathways (Berge et al., 2008).

Components of various medicinal plants used frequently in traditional Chinese medicine and in the Indian Ayurvedic system of medicine are claimed to have antiaging effects, which appear to be achieved through hormetic pathways. For example, celasterols and paeoniflorin present in some medicinal herbs used in Chinese medicine have cytoprotective effects and induce HSP in human cells (Westerheide et al., 2004; Yan et al., 2004). Similarly, curcumin, which is the active component in the commonly used yellow food spice from the roots of *Curcuma longa*, is a coinducer of HSP and has wide-ranging biological effects depending on its dose (Cronin, 2003; Dunsmore et al., 2001; Joe et al., 2004). Whereas curcumin doses greater than 10 μmole have been reported to have anti-inflammatory and anticancer effects in experimental studies (Moos et al., 2004; Rashmi et al., 2003), at lower doses (0.3 and 1 μmole) curcumin stimulates proteasome activity, enhances HSP induction after HS, and stimulates sodium pump activity (Ali and Rattan, 2006; Rattan and Ali, 2007). Several pharmaceuticals that also have typical hormetic dose-response effects (Calabrese, 2008) may be other examples of hormetins.

Hormesis may also be an explanation for the health beneficial effects of numerous other foods and food components, such as garlic, gingko, and other fruits and vegetables (Everitt et al., 2006; Ferrari, 2004; Gurib-Fakim, 2006; Hayes, 2005, 2007). Understanding the hormetic and interactive mode of action of natural and processed foods is a challenging field of research and has great potential for developing nutritional and other lifestyle modifications for aging intervention and therapies. For example, it may be possible to develop multihormetin formulations as antiaging drugs and nutriceuticals whose mode of action is through hormetic pathways by mild stress-induced stimulation of homeodynamic processes.

Other Stresses

Some examples of other stresses that have been investigated with respect to their effects in aging and longevity include starvation, electromagnetic stress, and mechanical stress, but the results are not consistent or well understood. For example, the effects of repeated physical injuries on lifespan have been studied for a marine oligochaete, *Paranais litoralis*, capable of posterior regeneration and of asexual reproduction (Martínez, 1996). Chronic low-frequency (10 Hz) electric stimulation of young and old male brown Norwegian rats resulted in more than twofold increase in the proportion of type IIa slow muscle fibers and in the content of satellite cells (Putman et al., 2001). Similarly, a long-wavelength, low-energy, and nonthermal electromagnetic frequency (50 MHz/0.5 W) enhanced cellular defenses of human T-cells and various aging characteristics in human fibroblasts (Perez et al., 2008). An example of low-level mechanical stress having beneficial hormetic effects was found in a study showing that a 20-minute burst of very low magnitude high-frequency vibrations given to the hind limbs of sheep increased the trabecular density by 34% in 1 year (Rubin et al., 2001). There is some indication that osteopontin synthesis in human dental osteoblasts is stimulated by low levels of mechanical stress (Liu et al., 2004).

Another kind of stress that appears to have hormetic effects is the population density in early stages of life. For example, it has been shown in *Drosophila* that larval crowding can induce both nutritional limitation and high concentrations of waste products and can thus be considered as a stressor for the larvae. Several studies reported that raising larvae in such conditions increased the lifespan of adult flies.

For instance, an increase in lifespan with increased larval density between 5 and 100 larvae per 5 cm^3 of food has been reported (Minois and Rattan, 2003). It has also been reported that whereas the developmental time, starvation resistance, relative fat content, and lifespan increased with larval density, viability was dramatically decreased from 91% to 50%–59% at density 350 (Minois and Rattan, 2003). The increase of the lifespan in those conditions might thus be due to a selection process at the larval stage. However, it has been shown that larval crowding without an effect on viability can increase lifespan in *D. melanogaster* (Sørensen and Loeschcke, 2001).

There are some studies attempting to check the hormetic effects of mental and psychological stress. Although the harmful effects of chronic and acute stress on life functioning, quality of life, and survival are well documented (Padgett and Glaser, 2003; Segerstrom and Miller, 2004), beneficial effects of periodic low-level mental stress are also being investigated. For example, C57BL6 young male mice exposed to stress by keeping them in a restrained space for 2.5 hours showed increased levels of stress hormone catecholamines and corticosterones and adrenal steroids and enhanced immunoprotection during surgery, vaccination, or infection (Viswanathan and Dhabhar, 2005). Skin fibroblasts from Cushing's syndrome patients, who have higher plasma levels of glucocorticoids, have longer replicative lifespan in vitro and have a better HSP stress response (Pratsinis et al., 2002; Zervolea et al., 2005). There are some preliminary studies that show that hormesis through mental challenge (Bierhaus et al., 2003) and through mind-concentrating meditational techniques may be useful in stimulating the stress response.

Hormesis Potential, Challenges, and Unresolved Issues in Aging

Because hormetic effects of mild stress are normally observed to be quite moderate, sometimes it is difficult to envisage the biological significance of hormesis in terms of its application in human aging intervention and prevention (Thayer et al., 2006; Zapponi and Marcello, 2006). However, it should be pointed out that although the initial hormetic effects may be relatively small when studied at the level of an individual biochemical step, often the final biological outcome, such as overall stress tolerance, functional improvement, and survival, is much larger, synergistic, and pleiotropic. This suggests that hormesis is involved in the biological amplification of adaptive responses leading to the improvement in overall cellular functions and performance. Exercise is a good example of the biological amplification of beneficial effects of mild stress, where not only do the specific muscle targets benefit, but also improvements in the immune system, cardiovascular system, sex hormones, libido, and mood are well documented. A recent study performed on rats showed that exercise performed at a young age can have lifelong benefits on bone structure and strength (Warden et al., 2007). This indicates that even moderate hormetic strengthening of homeodynamic networks can have much larger beneficial effects in terms of maintenance of functionality and prevention or delay of onset of age-related frailty.

The main promise and potential of hormesis as a modulator of aging lie in its mode of action. Because hormetic effects occur by involving a series of molecular and physiological processes, the final target of hormesis is the overall homeodynamic machinery of the living systems. Although hormesis-inducing stress may be targeted at a single pathway, the cascade of biological effects and their amplification result in the modulation and strengthening of the total homeodynamic ability. Furthermore, hormesis-induced increase in the prevention and removal of molecular damage will affect the rate of aging by slowing down the rate of shrinkage of the homeodynamic space and by reducing the increase in the size of the vulnerability zone (see Fig. 1).

As discussed earlier, the process of aging is primarily characterized by a progressive shrinking of homeodynamic space in terms of increased molecular heterogeneity, which leads to increased vulnerability, onset of diseases, and eventual death. It is also important to realize that the dimensions of the homeodynamic space of an individual are determined by an interacting network of genes, milieu, and chance, which are the basis of the uniqueness of the individual (Calabrese, 2008; Rattan, 2006). Several studies have been made and many are in progress to associate genetic variations (polymorphisms) with individual health status, probability of onset of various diseases, and lifespan potential (Christensen et al., 2006; Rattan and Singh, 2009). Because the practical applications of mild stress-induced hormesis are critically dependent on individual variations in stress response, studies to establish the association between stress gene variants and stress response are highly important and informative (Singh et al., 2004). Such studies are also necessary to establish the scientific foundations of so-called personalized medicine and personalized neutrigenomics (Calabrese, 2008; Dalton and Friend, 2006; Mutch et al., 2005).

Finally, there are some other important issues that remain to be resolved before hormesis can be successfully applied as an effective antiaging, health-promoting, and lifespan-extending strategy. Some of these issues are as follows:

- Establishing stress exposure regimens in terms of intensity and frequency
- Adjusting the levels of mild stress for age-related changes in the sensitivity to stress
- Establishing molecular criteria for identifying hormetic effects of different stresses
- Identifying qualitative and quantitative differences in stress response pathways initiated by different stressors
- Determining the interactive and pleiotropic effects of multiple stresses
- Determining the biological and evolutionary costs of repeated exposure to stress

In the context of modulating aging, repeated mild stress-induced hormesis increases the boundaries of the homeodynamic space, thus giving cells and organisms wider margins for metabolic fluctuation and adaptation. Slowing the shrinkage of the homeodynamic space hormetically will reduce the increase in the probability of occurrence and emergence of various diseases in old age and thus extend the health span.

References

Ali RE, Rattan SIS (2006) Curcumin's biphasic hormetic response on proteasome activity and heat shock protein synthesis in human keratinocytes. Ann N Y Acad Sci 1067: 394–399.

Anson RM, Guo Z, de Cabo R, Lyun T, Rios M, Hagepanos A, Ingram DK, Lane MA, Mattson MP (2003) Intermittent fasting dissociates beneficial effects of dietary restriction on glucose metabolism and neuronal resistance to injury from calorie restriction. Proc Natl Acad Sci USA 100: 6216–6220.

Arumugum TV, Gleichmann M, Tang SC, Mattson MP (2006) Hormesis/preconditioning mechanisms, the nervous system and aging. Age Res Rev 5: 165–178.

Atalay M, Oksala NK, Laaksonen DE, Khanna S, Nakao C, Lappalainen J, Roy S, Hanninen OO, Sen CK (2004) Exercise training modulates heat shock protein response in diabetic rats. J Appl Physiol 97: 605–611.

Atkinson WD, Law DV, Bromley KJ, Inskip HM (2004) Mortality of employees of the United Kingdom Atomic Energy Authority, 1946–97. Occup Environ Med 61: 577–585.

Baur JA, Pearson KJ, Price NL, Jamieson HA, Lerin C, Kalra A, Prabhu VK, Allard JS, Lopez-Lluch G, Lewis K, et al. (2006) Resveratrol improves health and survival of mice on a high-calorie diet. Nature 443: 337–342.

Beedholm R, Clark BFC, Rattan SIS (2004) Mild heat stress stimulates proteasome and its 11s activator in human fibroblasts undergoing aging in vitro. Cell Stress Chaperones 9: 49–57.

Berge U, Kristensen P, Rattan SIS (2008) Hormetic modulation of differentiation of normal human epidermal keratinocytes undergoing replicative senescence in vitro. Exp Gerontol 43: 658–662.

Bierhaus A, Wolf J, Andrassy M, Rohleder N, Humpert PM, Petrov D, Fersti R, von Eynatten M, Wendt T, Rudofsky G, et al. (2003) A mechanism converting psychosocial stress into mononuclear cell activation. Proc Natl Acad Sci USA 100: 1920–1925.

Bonelli MA, Desenzani S, Cavallini G, Donati A, Romani AA, Bergamini E, Borghetti AF (2008) Low-level caloric restriction rescues proteasome activity and hsc70 level in liver of aged rats. Biogerontology 9: 1–10.

Butov A, Johnson TE, Cypser J, Sannikov I, Volkov M, Sehl M, Yashin AI (2001) Hormesis and debilitation effects in stress experiments using the nematode worm *Caenorhabditis elegans:* the model of balance between cell damage and HSP level. Exp Gerontol 37: 57–66.

Calabrese EJ (2004) Hormesis: from marginalization to mainstream: a case for hormesis as the default dose-response model in risk assessment. Toxicol Appl Pharmacol 197: 125–136.

Calabrese EJ (2008) Hormesis and medicine. Br J Clin Pharmacol 66: 594–617.

Calabrese EJ, Baldwin LA (2000) The effects of gamma rays on longevity. Biogerontology 1: 309–319.

Caratero A, Courtade M, Bonnet L, Planel H, Caratero C (1998) Effect of continuous gamma irradiation at a very low dose on the life span of mice. Gerontology 44: 272–276.

Cardis E, Vrijheid M, Blettner M, Gilbert E, Hakama M, Hill C, Howe G, Kaldor J, Muirhead CR, Schubauer-Berigan M, et al. (2007) The 15-country collaborative study of cancer risk among radiation workers in the nuclear industry: estimates of radiation-related cancer risks. Radiat Res 167: 396–416.

Christensen K, Johnson TE, Vaupel JW (2006) The quest for genetic determinants of human longevity: challenges and insights. Nat Rev Genet 7: 436–448.

Cologne JB, Preston DL (2000) Longevity of atomic-bomb survivors. Lancet 356: 303–307.

Corder R, Mullen W, Khan NQ, Marks SC, Wood EG, Carrier MJ, Crozier A (2006) Red wine procyanidins and vascular health. Nature 444: 566.

Cronin JR (2003). Curcumin: old spice is a new medicine. Alternative Complement Ther 2003(Feb): 34–38.

Cypser JR, Johnson TE (2002) Multiple stressors in *Caenorhabditis elegans* induce stress hormesis and extended longevity. J Gerontol Biol Sci 57A: B109–B114.

Cypser JR, Johnson TE (2003) Hormesis in *Caenorhabditis elegans* dauer-defective mutants. Biogerontology 4: 203–214.

Dalton WS, Friend SH (2006) Cancer biomarkers—an invitation to the table. Science 312: 1165–1168.

Dunsmore KE, Chen PG, Wong HR (2001) Curcumin, a medicinal herbal compound capable of inducing heat shock response. Crit Care Med 29: 2199–2204.

Eller MS, Maeda T, Magnoni C, Atwal D, Gilchrest BA (1997) Enhancement of DNA repair in human skin cells by thymidine dinucleotides: evidence for a p53-mediated mammalian SOS response. Proc Natl Acad Sci USA 94: 12627–12632.

Everitt AV, Hilmer SN, Brand-Miller JC, Jamieson HA, Truswell AS, Sharma AP, Mason RS, Morris BJ, Le Couteur DG (2006) Dietary approaches that delay age-related diseases. Clin Interv Aging 1: 11–31.

Ferrari CKB (2004) Functional foods, herbs and neutraceuticals: towards biochemical mechanisms of healthy aging. Biogerontology 5: 275–289.

Fonager J, Beedholm R, Clark BFC, Rattan SIS (2002) Mild stress-induced stimulation of heat shock protein synthesis and improved functional ability of human fibroblasts undergoing aging in vitro. Exp Gerontol 37: 1223–1238.

Fontana L, Meyer TE, Klein S, Holloszy JO (2004) Long-term calorie restriction is highly effective in reducing the risk for atherosclerosis in humans. Proc Natl Acad Sci USA 10: 6659–6663.

Gomez J, Caro P, Naudi A, Portero-Otin M, Pamplona R, Barja G (2007) Effect of 8.5% and 25% caloric restriction on mitochondrial free radical production and oxidative stress in rat liver. Biogerontology 8: 555–566.

Goukassian DA, Helms E, Van Steeg H, van Oostrom C, Bhawan J, Gilchrest BA (2004) Topical DNA oligonucleotide therapy reduces UV-induced mutations and photocarcinogenesis in hairless mice. Proc Natl Acad Sci USA 101: 3933–3938.

Gurib-Fakim A (2006) Medicinal plants: traditions of yesterday and drugs of tomorrow. Mol Asp Med 27: 1–93.

Hayakawa N, Ohtaki M, Ueoka H, Matsuura M, Munaka M, Kurihara M (1989) Mortality statistics of major causes of death among atomic bomb survivors in Hiroshima Prefecture from 1968 to 1982. Hiroshima J Med Sci 38: 53–67.

Hayes DP (2005) The protective role of fruits and vegetables against radiation-induced cancer. Nutr Rev 63: 303–311.

Hayes DP (2007) Nutritional hormesis. Eur J Clin Nutr 61: 147–159.

Hayflick L (2007) Biological aging is no longer an unsolved problem. Ann N Y Acad Sci 1100: 1–13.

Heilbronn LK, de Jonge L, Frisard, M.I., DeLany, J.P., Larson-Meyer, D.E., Rood J., Nguyen, T., Martin, C.K., Volaufova, J., Most, M.M., Greenway, FL, et al (2006) Effect of 6-month calorie restriction on biomarkers of longevity, metabolic adaptation, and oxidative stress in overweight individuals. JAMA 295: 1539–1548.

Hercus MJ, Loeschcke V, Rattan SIS (2003) Lifespan extension of *Drosophila melanogaster* through hormesis by repeated mild heat stress. Biogerontology 4: 149–156.

Holliday R (1991) A re-examination of the effects of ionizing radiation on lifespan and transformation of human diploid fibroblasts. Mutat Res 256: 295–302.

Holliday R (2006) Aging is no longer an unsolved problem in biology. Ann N Y Acad Sci. 1067: 1–9.

Holliday R (2007) Ageing: the paradox of life. Dordrecht, The Netherlands: Springer.

Holloszy JO, Smith EK (1986) Longevity of cold-exposed rats: a reevaluation of the rate of living theory. J Appl Physiol 61: 1656–1660.

Ina Y, Sakai K (2005) Further study of prolongation of life span associated with immunologic modification by chronic low-dose-rate irradiation in MRL-lpr/lpr mice: effects of whole-life irradiation. Radiat Res 163: 418–423.

Ina Y, Tanooka H, Yamada T, Sakai K (2005) Suppression of thymic lymphoma induction by life-long low-dose-rate irradiation accompanied by immune activation in C57BL/6 mice. Radiat Res.163: 153–158.

Ji LL (2008) Physical activity: a strong stimulant for hormesis during aging. In: Mild stress and healthy aging: applying hormesis in aging research and interventions, Le Bourg E, Rattan SIS, eds. Dordrecht, The Netherlands: Springer, pp. 99–114.

Joe B, Vijaykumar M, Lokesh BR (2004) Biological properties of curcumin—cellular and molecular mechanism of action. Crit Rev Food Sci Nutr 44: 97–111.

Johnson TE (2002) A personal retrospective on the genetics of aging. Biogerontology 3: 7–12.

Johnson TE, Hartman PS (1988) Radiation effects on life span in *Caenorhabditis elegans*. J Gerontol 43: B137–B141.

Keany M, Matthijssens F, Sharpe M, Vanfleteren J, Gems D (2004) Superoxide dismutase mimetics elevate superoxide dismutase activity in vivo but do not retard aging in the nematode *Caenorhabditis elegans*. Free Radic Biol Med 37: 239–250.

Khaw KT, Wareham N, Bingham S, Welch A, Luben R, Day N (2008) Combined impact of health behaviours and mortality in men and women: the EPIC-Norfolk Prospective Population Study. PLoS Med 5: e12.

Khazaeli AA, Tatar M, Pletcher SD, Curtsinger JW (1997) Heat-induced longevity extension in *Drosophila*. I. Heat treatment, mortality, and thermotolerance. J Gerontol Biol Sci 52A: B48–B52.

Kortlever RM, Bernards R (2006) Senescence, wound healing and cancer: the PAI-1 connection. Cell Cycle 5: 2697–2703.

Kraft DC, Deocaris CC, Rattan SIS (2006) Proteasomal oscillation during mild heat shock in aging human skin fibroblasts. Ann N Y Acad Sci 1067: 224–227.

Lamming DW, Wood JG, Sinclair DA (2004) Small molecules that regulate lifespan: evidence for xenohormesis. Mol Microbiol 53: 1003–1009.

Lancaster GI, Møller K, Nielsen B, Secher NH, Febbraio MA, Nybo L (2004) Exercise induces the release of heat shock protein 72 from the human brain in vivo. Cell Stress Chaperones 9: 276–280.

Le Bourg E (2008) Hypergravity in drosophila melanogaster. In: Mild stress and healthy aging: applying hormesis in aging research and interventions, Le Bourg E, Rattan SIS, eds. Dordrecht, The Netherlands: Springer, pp. 63–63.

Le Bourg E, Valenti P, Lucchetta P, Payre F (2001) Effects of mild heat shocks at young age on aging and longevity in *Drosophila melanogaster*. Biogerontology 2: 155–164.

Linnane AW, Eastwood H (2006) Cellular redox regulation and prooxidant signaling systems. A new perspective on the free radical theory of aging. Ann N Y Acad Sci 1067: 47–55.

Lithgow GJ, White TM, Melov S, Johnson TE (1995) Thermotolerance and extended life-span conferred by single-gene mutations and induced by thermal stress. Proc Natl Acad Sci USA 92: 7540–7544.

Liu R, Liu IY, Bi X, Thompson RF, Doctrow SR, Malfroy B, Baudry M (2003) Reversal of age-related learning deficits and brain oxidative stress in mice with superoxide dismutase/catalase mimetics. Proc Natl Acad Sci USA 100: 8526–8531.

Liu D, Vandahl BB, Birkelund S, Nielsen LB, Melsen B (2004) Secretion of osteopontin from MG-63 cells under a physiological level of mechanical strain in vitro—a [35s] incorporation approach. Eur J Orthodont 26: 143–149.

Martin B, Mattson MP, Maudsley S (2006) Caloric restriction and intermittent fasting: two potential diets for successful aging. Age Res Rev 5: 332–353.

Martínez DE (1996) Rejuvenation of the disposable soma: repeated injury extends lifespan in an asexual annelid. Exp Gerontol 31: 699–704.

Masoro EJ (2007) The role of hormesis in life extension by dietary restriction. Interdiscip Top Gerontol 35: 1–17.

Mattson MP (2008) Dietary factors, hormesis and health. Ageing Res Rev 7: 43–48.

Melov S, Ravenscroft J, Malik S, Gill MS, Walker DW, Clayton PE, Wallace DC, Malfroy B, Doctrow SR, Lithgow GJ (2000) Extension of life-span with superoxide dismutase/catalase mimetics. Science 289: 1567–1569.

Meyer TE, Kovács SK, Ehsani AA, Klein S, Holloszy JO, Fontana L (2006) Long-term caloric restriction ameliorates the decline in diastolic functions in humans. J Am Coll Cardiol 47: 398–402.

Michalski AI, Johnson TE, Cypser JR, Yashin AI (2001) Heating stress patterns in *Caenorhabditis elegans* longevity and survivorship. Biogerontology 2: 35–44.

Mine M, Okumura Y, Ichimaru M, Nakamura T, Kondo S (1990) Apparently beneficial effect of low to intermediate doses of A-bomb radiation on human lifespan. Int J Radiat Biol 58: 1035–1043.

Minois N (2000) Longevity and aging: beneficial effects of exposure to mild stress. Biogerontology 1: 15–29.

Minois N (2006) The hormetic effects of hypergravity on longevity and aging. Dose-response 4: 123–132.

Minois N, Khazaeli AA, Curtsinger JW (2001) Locomotor activity as a function of age and life span in *Drosophila melanogaster* overexpressing hsp70. Exp Gerontol 36: 1137–1153.

Minois N, Rattan SIS (2003) Hormesis in aging and longevity. In: Modulating aging and longevity, Rattan SIS, ed. Dordrecht, The Netherlands: Kluwer Academic Publishers, pp. 127–137.

Minois N, Vaynberg S (2002) Fecundity and life span in transgenic *Drosophila melanogaster* overexpressing hsp70. Biogerontology 3: 301–306.

Mocchegiani E, Rink L, Blasco MA (2006) Zinc and ageing (ZINCAGE Project). Biogerontology 7: 305–306.

Moore MN (2008) Autophagy as a second level protective process in conferring resistance to environmentally-induced oxidative stress. Autophagy 4: 254–256.

Moos PJ, Edes K, Mullally JE, Fitzpatrick FA (2004) Curcumin impairs tumour suppressor p53 function in colon cancer cells. Carcinogenesis 25: 1611–1617.

Mutch DM, Wahli W, Williamson G (2005) Nutrigenomics and nutrigenetics: the emerging faces of nutrition. FASEB J 19: 1602–1616.

Nambi KS, Soman SD (1987) Environmental radiation and cancer in India. Health Phys 52: 653–657.

Nielsen ER, Eskildsen-Helmond Y, Rattan SIS (2006) MAP-kinases and heat shock–induced hormesis in human fibroblasts during serial passaging in vitro. Ann N Y Acad Sci 1067: 343–348.

Nørgaard R, Kassem M, Rattan SIS (2006) Heat shock–induced enhancement of osteoblastic differentiation of hTERT-immortalized mesenchymal stem cells. Ann N Y Acad Sci 1067: 443–447.

Okajima S, Mine M, Nakamura T (1985) Mortality of registered A-bomb survivors in Nagasaki, Japan, 1970–1984. Radiat Res 103: 419–431.

Olsen A, Vantipalli MC, Lithgow GJ (2006) Lifespan extension of *Caenorhabditis elegans* following repeated mild hormetic heat treatments. Biogerontology 7: 221–230.

Ordy JM, Samorajski T, Zeman W, Curtis HJ (1967) Interaction effects of environmental stress and deuteron irradiation of the brain on mortality and longevity of C57BL/10 mice. Proc Soc Exp Biol Med 126: 184–190.

Overgaard J, Sørensen JG, Petersen SO, Loeschcke V, Holmstrup M (2005) Changes in membrane lipid composition following rapid cold hardening in *Drosophila melanogaster*. J Insect Physiol 51: 1173–1182.

Padgett RW, Glaser R (2003) How stress influences the immune response. Trends Immunol 24: 444–448.

Park HG, Han SI, Oh SY, Kang HS (2005) Cellular responses to mild heat stress. Cell Mol Life Sci 62: 10–23.

Parsons PA (2003) Energy, stress and the invalid linear no-threshold premise: a generalization illustrated by ionizing radiation. Biogerontology 4: 227–231.

Perez FP, Zhou X, Morisaki J, Jurivich D (2008) Electromagnetic field therapy delays cellular senescence and death by enhancement of the heat shock response. Exp Gerontol 43: 307–316.

Pratsinis H, Tsagarakis S, Zervolea I, Giannakopoulos F, Stathakos D, Thalassinos N, Kletsas D (2002) Chronic in vivo exposure to glucocorticoids prolongs cellular lifespan: the case of Cushing's syndrome–patients′ fibroblasts. Exp Gerontol 37: 1237–1245.

Putics A, Végh EM, Csermely P, Soti C (2008) Resveratrol induces the heat-shock response and protects human cells from severe heat stress. Antiox Red Sign 10: 1–11.

Putman CT, Sultan KR, Wassmer T, Bamford JA, Skorjanc D, Pette D (2001) Fiber-type transitions and satellite cell activation in low-frequency-stimulated muscles of young and aging rats. J Gerontol Biol Sci 56A: B510–B519.

Radak Z, Chung HY, Goto S (2005) Exercise and hormesis: oxidative stress-related adaptation for successful aging. Biogerontology 6: 71–75.

Radak Z, Chung HY, Goto S (2008a) Systemic adaptation to oxidative challenge induced by regular exercise. Free Radic Biol Med 44: 153–159.

Radak Z, Chung HY, Koltai E, Taylor AW, Goto S (2008b) Exercise, oxidative stress and hormesis. Ageing Res Rev 7: 34–42.

Raji NS, Surekha A, Subba Rao K (1998) Improved DNA-repair parameters in PHA-stimulated peripheral blood lymphocytes of human subjects with low body mass index. Mech Ageing Dev 104: 133–148.

Rashmi R, Santhosh Kumar TR, Karunagaran D (2003) Human colon cancer cells differ in their sensitivity to curcumin-induced apoptosis and heat shock protects them by inhibiting the release of apoptosis-inducing factor and caspases. FEBS Lett 538: 19–24.

Rattan SIS (1998) Repeated mild heat shock delays ageing in cultured human skin fibroblasts. Biochem Mol Biol Int 45: 753–759.

Rattan SIS (2000a) Ageing, gerontogenes, and hormesis. Ind J Exp Biol 38: 1–5.

Rattan SIS (2000b) Biogerontology: the next step. Ann N Y Acad Sci 908: 282–290.

Rattan SIS (2001a) Applying hormesis in aging research and therapy. Hum Exp Toxicol 20: 281–285.

Rattan SIS (2001b) Hormesis in biogerontology. Crit Rev Toxicol 31: 663–664.

Rattan SIS (2004) Aging intervention, prevention, and therapy through hormesis. J Gerontol Biol Sci 59A: 705–709.

Rattan SIS (2005) Anti-ageing strategies: prevention or therapy? EMBO Rep 6: S25–S29.

Rattan SIS (2006) Theories of biological aging: genes, proteins and free radicals. Free Radic Res 40: 1230–1238.

Rattan SIS (2007a) Homeostasis, homeodynamics, and aging. In: Encyclopedia of gerontology, Birren J, ed. London: Elsevier, pp. 696–699.

Rattan SIS (2007b) The science of healthy aging: genes, milieu, and chance. Ann N Y Acad Sci 1114: 1–10.

Rattan SIS (2008) Hormesis in aging. Ageing Res Rev 7: 63–78.

Rattan SIS, Ali RE (2007) Hormetic prevention of molecular damage during cellular aging of human skin fibroblasts and keratinocytes. Ann N Y Acad Sci 1100: 424–430.

Rattan SIS, Clark BFC (2005) Understanding and modulating ageing. IUBMB Life 57: 297–304.

Rattan SIS, Fernandes RA, Demirovic D, Dymek B, Lima CF (2009) Heat stress and hormetin-induced hormesis in human cells: effects on aging, wound healing, angiogenesis and differentiation. Dose-response 7: 90–103.

Rattan SIS, Singh R (2009) Gene therapy in ageing. Gene Therapy 16: 3–9.

Ritzmann RF, Glasky A, Steinberg A, Melchior CL (1994) The interaction of ethanol with the cognitive enhancers tacrine, physostigmine, and AIT-082. J Gerontol 49: B51–B53.

Rogina B, Helfand SL (2004) Sir2 mediates longevity in the fly through a pathway related to calorie restriction. Proc Natl Acad Sci USA 101: 15998–16003.

Rubin C, Turner AS, Bain S, Mallinckrodt C, McLeod K (2001) Low mechanical signals strengthen long bones. Nature 412: 603–604.

Safwat A (2008) Clinical applications of low-dose whole body irradiation hormesis. In: Mild stress and healthy aging: applying hormesis in aging research and interventions, Le Bourg E, Rattan SIS, eds. Dordrecht, The Netherlands: Springer, pp. 157–170.

Segerstrom SC, Miller GE (2004) Psychological stress and the human immune system: a meta-analytic study of 30 years of inquiry. Psychol Bulletin 130: 601–630.

Sharma S, Kaur G (2005) Neuroprotective potential of dietary restriction against kainate-induced excitotoxicity in adult male Wistar rats. Brain Res Bull 67: 482–491.

Short KR, Vittone JL, Bigelow ML, Proctor DN, Nair KS (2004) Age and aerobic exercise training effects on whole body muscle protein metabolism. Am J Physiol Endocrinol Metab 286: E92–E101.

Singh R, Kølvraa S, Bross P, Gregersen N, Nexø BA, Frederiksen H, Christensen K, Rattan SIS (2004) Association between low self-rated health and heterozygosity for -110A > C polymorphism in the promoter region of HSP70-1 in aged Danish twins. Biogerontology 5: 169–176.

Sogawa H, Kubo C (2000) Influence of short-term repeated fasting on the longevity of female (NZB x NZW)F1 mice. Mech Ageing Dev 115: 61–71.

Sørensen JG, Loeschcke V (2001) Larval crowding in *Drosophila melanogaster* induces hsp70 expression, and leads to increased adult longevity and adult thermal stress resistance. J Insect Physiol 47: 1301–1307.

Sørensen JG, Sarup P, Kristensen TN, Loeschcke V (2008) Temperature-induced hormesis in *Drosophila*. In: Mild stress and healthy aging: applying hormesis in aging research and interventions, Le Bourg E, Rattan SIS, eds. Dordrecht, The Netherlands: Springer, pp. 65–80.

Sun J, Liao JK (2004) Induction of angiogenesis by heat shock protein 90 mediated by protein kinase akt and endothelial nitric oxide synthase. Arterioscler Thromb Vasc Biol 24: 2238–2244.

Sun Y, MacRae TH (2005) The small heat shock proteins and their role in human disease. FEBS J 272: 2613–2627.

Suzuki K, Kodama S, Watanabe M (1998a) Suppressive effects of low-dose preirradiation on genetic instability induced by X rays in normal embryonic cells. Radiat Res 150: 656–662.

Suzuki K, Kodama S, Watanabe M (2001) Extremely low-dose ionizing radiation causes activation of mitogen-activated protein kinase pathway and enhances proliferation of normal human diploid cells. Cancer Res 61: 5396–5401.

Suzuki M, Yang Z, Nakano K, Yatagai F, Suzuki K, Kodama S, Watanabe M (1998b) Extension of in vitro life-span of gamma-irradiated human embryo cells accompanied by chromosome instability. J Radiat Res 39: 203–213.

Thayer KA, Melnick R, Huff J, Burns K, Davis D (2006) Hormesis: a new religion? Environ Health Perspect 114: A632–633.

Tsutsui T, Tanaka Y, Matsudo Y, Hasegawa K, Fujino T, Kodama S, Barrett JC (1997) Extended lifespan and immortalization of human fibroblasts induced by X-ray irradiation. Mol Carcinog 18: 7–18.

Vaiserman AM (2008) Irradiation and hormesis. In: Mild stress and healthy aging: applying hormesis in aging research and interventions, Le Bourg E, Rattan SIS, eds. Dordrecht, The Netherlands: Springer, pp. 21–41.

Vaiserman AM, Koshel NM, Litoshenko AY, Mozzhukhina TG, Voitenko VP (2003) Effects of X-irradiation in early ontogenesis on the longevity and amount of the S1 nuclease-sensitive DNA sites in adult *Drosophila melanogaster*. Biogerontology 4: 9–14.

Vaiserman AM, Koshel NM, Mechova LV, Voitenko VP (2004a) Cross-life stage and cross-generational effects of g irradiation at the egg stage on *Drosophila melanogaster* life histories. Biogerontology 5: 327–337.

Vaiserman AM, Koshel NM, Voitenko VP (2004b) Effect of X-irradiation at larval stage on adult lifespan in *Drosophila melanogaster*. Biogerontology 5: 49–54.

Valenzano DR, Terzibasi E, Genade T, Cattaneo A, Domenici L, Cellerino A (2006) Resveratrol prolongs lifespan and retards the onset of age-related markers in a short-lived vertebrate. Curr Biol 16: 296–300.

Verbeke P, Clark BFC, Rattan SIS (2001a) Reduced levels of oxidized and glycoxidized proteins in human fibroblasts exposed to repeated mild heat shock during serial passaging in vitro. Free Radic Biol Med 31: 1593–1602.

Verbeke P, Deries M, Clark BFC, Rattan SIS (2002) Hormetic action of mild heat stress decreases the inducibility of protein oxidation and glycoxidation in human fibroblasts. Biogerontology 3: 105–108.

Verbeke P, Fonager J, Clark BFC, Rattan SIS (2001b) Heat shock response and ageing: mechanisms and applications. Cell Biol Int 25: 845–857.

Viswanathan K, Dhabhar FS (2005) Stress-induced enhancement of leukocyte trafficking into sites of surgery or immune activation. Proc Natl Acad Sci USA 102: 5808–5813.

Vrijheid M, Cardis E, Ashmore P, Auvinen A, Bae JM, Engels H, Gilbert E, Gulis G, Habib R, Howe G, et al. (2007) Mortality from diseases other than cancer following low doses of ionizing radiation: results from the 15-country study of nuclear industry workers. Int J Epidemiol 36: 1126–1135.

Walford RL, Mock D, Verdery R, MacCallum T (2002) Calorie restriction in biosphere 2: alterations in physiologic, hematologic, hormonal, and biochemical parameters in humans restricted for a 2-year period. J Gerontol Biol Sci 57A: B211–B224.

Warden SJ, Fuchs RK, Castillo AB, Nelson IR, Turner CH (2007) Exercise when young provides lifelong benefits to bone structure and strength. J Bone Miner Res 22: 251–259.

Watanabe M, Suzuki M, Suzuki K, Nakano K, Watanabe K (1992) Effect of multiple irradiation with low doses of gamma-rays on morphological transformation and growth ability of human embryo cells in vitro. Int J Radiat Biol 62: 711–718.

Westerheide SD, Bosman JD, Mbadugha BNA, Kawahara TLA, Matsumoto G, Kim S, Gu W, Devlin JP, Silverman RB, Morimoto RI (2004) Celastrols as inducers of the heat shock response and cytoprotection. J Biol Chem 279: 56053–56060.

Wood JG, Rogina B, Lavu S, Howitz KT, Helfand SL, Tatar M, Sinclair DA (2004) Sirtuin activators mimic caloric restriction and delay ageing in metazoans. Nature 430: 686–689.

Wyngaarden KEV, Pauwels EKJ (1995) Hormesis: are low doses of ionizing radiation harmful or beneficial? Eur J Nucl Med 22: 481–486.

Yan D, Saito K, Ohmi Y, Fujie N, Ohtsuka K (2004) Paeoniflorin, a novel heat shock protein-inducing compound. Cell Stress Chaperones 9: 378–389.

Yashin AI, Cypser JR, Johnson TE, Michalski AI, Boyko SI, Novoseltsev VN (2001) Ageing and survival after different doses of heat shock: the results of analysis of data from stress experiments with the nematode worm *Caenorhabditis elegans*. Mech Ageing Dev 122: 1477–1495.

Yates FE (1994) Order and complexity in dynamical systems: homeodynamics as a generalized mechanics for biology. Math Comput Model 19: 49–74.

Yokoyama K, Fukumoto K, Murakami T, Harada S, Hosono R, Wadhwa R, Mitsui Y, Ohkuma S (2002) Extended longevity of *Caenorhabditis elegans* by knocking in extra copies of hsp70f, a homolog of mot-2 (mortalin)/mthsp70/grp75. FEBS Lett 516: 53–57.

Zapponi GA, Marcello I (2006) Low-dose risk, hormesis, analogical and logical thinking. Ann N Y Acad Sci 1076: 839–857.

Zervolea I, Pratsinis H, Tsagarakis S, Karavitaki N, Stathakos D, Thalassinos N, Kletsas D (2005) The impact of chronic in vivo glucocorticoid excess on the functional characteristics of human skin fibroblasts obtained from patients with endogeneous Cushing's syndrome. Eur J Endocrinol 152: 895–902.

The Hormetic Pharmacy: The Future of Natural Products and Man-Made Drugs in Disease Prevention and Treatment

Edward J. Calabrese and Mark P. Mattson

Abstract This chapter proposes a new look at the modern pharmacy of natural products and man-made drugs. It shows that local pharmacies have been dispensing drugs based on hormesis for many decades. Among drugs that act via the hormetic dose response are those that combat anxiety and depression, make bones stronger, grow hair thicker, enhance cognitive function, and lessen pain. These drugs are some of the staples of the industry, permitting millions of people to live more normal and better lives. This chapter also shows that even when commonly used drugs such as antibiotics are administered to kill bacteria, they act hormetically at low doses and may be of concern because they can cause harmful bacterial colonies to grow. This is the case for antitumor drugs as well. Thus, the modern pharmacy is really a hormetic pharmacy, but this is really just the beginning. The currently unfolding future is one in which the biomedical and pharmaceutical giants will be developing so-called hormetic mimetics—drugs that can induce the normal adaptive pathways seen in the thousands of studies demonstrating hormesis but without exposing individuals to toxic doses of chemicals, ionizing radiation, extreme heat, or hard-to-maintain caloric restriction regimens. Screening chemicals based on their ability to activate, at subtoxic doses, specific adaptive cellular stress response pathways is a promising approach for drug development. The future pharmacy will be enhanced by research strategies and clinical practices that adopt hormetic principles and applications.

Keywords Antidepressants · Cardiovascular disease · Dietary supplements · Dose and frequency · Nutriceuticals · Pharmaceutical industry

E.J. Calabrese (✉)
Department of Environmental Health Sciences Division, School of Public Health and Health Sciences, University of Massachusetts, Amherst, MA 01003, USA
e-mail: edwardc@schoolph.umass.edu

M.P. Mattson, E.J. Calabrese, *Hormesis*, DOI 10.1007/978-1-60761-495-1_10,

Introducing the Pharmacist to Hormesis

The concept of hormesis is a central and integrative feature of modern pharmacology and the pharmaceutical industry. This may seem odd, given that neither entity has used the term very often, if at all. Yet entire areas of drug development (e.g., anxiolytic/antianxiety drugs, antiseizure drugs, cognitive/memory-enhancing drugs, male sexual dysfunction/erectile dysfunction drugs), as well as many other drugs on the market, are based to a considerable degree on the hormetic dose response. In fact, we would contend that the history of drug development, although empirically making use of the concept of hormesis and building an industry with it, has not been guided by its underlying theory and its numerous practical applications, thereby retarding the successful development of marketable drugs. We believe that the field of pharmacology and the pharmaceutical industry would benefit greatly from the use of hormesis-based strategies in drug discovery, development, and therapeutic applications within clinical settings. Hormesis-based concepts and approaches also extend to the regulatory community, such as the U.S. Food and Drug Administration (FDA), which could take a leadership role to provide direction in this critical area.

This chapter assesses how the pharmaceutical and natural product industries have been using hormesis in the development of numerous products but without appreciating that the hormesis concept has been a driving factor in drug development and therapeutic dosing. It also shows how this lack of understanding impedes seeing hormesis as a general biological concept that could be applied to other critical areas of the pharmaceutical industry. It addresses how hormesis affects key pharmacological concepts such as the dose response and its quantitative features, understandings of drug potency within a dose-response context, the width of the therapeutic zone, drug interactions, and the design of preclinical and clinical studies. A general overview of the occurrence of hormesis within a broad range of pharmaceutical target areas is presented. A final section is directed toward the development of agents that can activate hormetic pathways (see the chapters Hormesis: What It Is and Why It Matters and The Devil Is in the Dose: Complexity of Receptor Systems and Responses for examples of specific hormetic signaling pathways), resulting in protection against a plethora of human diseases and age-related conditions. In effect, this chapter introduces the reader to the emerging world of the hormetic pharmacy.

Why the Pharmaceutical Industry Is Missing the Hormesis Revolution

The nature of research in the biomedical/pharmacological sciences has become progressively and strikingly specialized. This is seen with the development of a bewildering array of professional societies and their subsequent splintering and eventual subdividing into further, more focused specialties. Bates (1965) provided a detailed history of scientific societies throughout a large portion of the 20th century. This hyperspecialization process has been widespread and has led to profound advances in all areas of scientific and engineering research. However, such specialization is not without it limitations or concerns. In most cases, each professional society develops its own means of communication, educational and technology plans, professional meetings, terminology, concepts, priorities, and agenda. Once closely related professional societies, no longer regularly communicating, have even developed different names for the same concept. This is seen in the area of the dose response, where approximately two dozen terms, most often discipline specific, have been used to describe the hormetic dose-response relationship (Calabrese, 2008a, b). In general, there is a strong tendency to only follow developments in one's narrow area of research and teaching interest. This tendency has had a major impact on the recognition of the hormetic dose response and its generalization and applications to other areas. In effect, the general nature of the hormetic dose response became lost or missed within this process of hyperspecialization as various biomedical subdisciplines failed to appreciate the general nature of the hormetic biphasic dose-response relationship and its broad implications. However, the resurgence of interest in the area of hormesis has revealed striking and important commonalities with respect to dose-response relationships across the various biological and biomedical disciplines. This comparative assessment of biphasic dose-response relationships has revealed similarities with respect to temporal and quantitative features and mechanistic strategies all set within a framework of biological plasticity. Such assessments have revealed that most, if not all, drugs act within a hormetic framework. This finding supports the conclusion that the strong tendency toward specialization has tended to prevent the discovery of new processes based on the biological concept that essentially all cell types, tissues, and organisms follow an evolution-based hormetic dose-response strategy (see also chapter The Fundamental Role of Hormesis in Evolution).

Hormesis and Biological Plasticity

Dietary Factors

Several dietary factors that have been conclusively shown to affect health and modify the aging process exhibit biphasic dose responses. We believe that low doses of such dietary factors may exert their beneficial effects by imposing a mild stress on cells that is analogous to that of physical exercise. Regular light to moderate exercise decreases the risk of several major diseases, including cardiovascular disease, stroke, diabetes, and Alzheimer's disease (see the chapter Exercise-Induced Hormesis). In muscle cells and apparently in other cells, including those in the nervous and reproductive systems (van Praag, 2008; Chigurupati et al., 2008), exercise results in metabolic and oxidative stress. As long as the intensity and duration are not too extensive, the cells respond adaptively to the exercise period by increasing the production of proteins that increase their resistance to more severe stress and protect against degeneration and disease (Arumugam et al., 2006; Radak et al.,

2008). However, sustained high-intensity exercise can result in tissue damage and even death (Hubbard, 1990; Drory et al., 1991).

One dietary factor that appears to tap into some of the same adaptive stress response pathways is moderate dietary energy restriction. Studies of humans have shown that excessive energy intake increases the risk for a range of major diseases, including diabetes, cardiovascular and cerebrovascular diseases, and several different types of cancer (Pender and Pories, 2005). Recent findings suggest that dietary energy restriction improves cardiovascular risk profiles (Fontana, 2008; Redman et al., 2008) and can reverse oxidative and inflammatory processes underlying disease processes, including asthma (Johnson et al., 2007). In rodents moderate reductions in energy intake (10% to 40%) and/or intermittent fasting increase lifespan and reduce the incidence of disorders such as diabetes, cancers, and sarcopenia (Carter et al., 2007; Colman et al., 2008). We and others have provided evidence that a major component of the mechanism by which moderate dietary energy restriction counteracts aging and protects cells against disease is by inducing a hormetic response (see the chapter Dietary Energy Intake, Hormesis, and Health). For example, dietary energy restriction results in increased production of neuroprotective proteins in brain cells, including brain-derived neurotrophic factor (BDNF), glial cell line–derived neurotrophic factor (GDNF), heat-shock proteins, and mitochondrial uncoupling protein 4 (Yu and Mattson, 1999; Lee et al., 2002; Maswood et al., 2004; Liu et al., 2006). Conversely, excessive energy intake may impair adaptive cellular stress response pathways (see the chapter Hormesis and Aging).

Plants produce a remarkable array of chemicals and also contain numerous chemicals produced by bacteria or fungi that live within (endophytes) or on (epiphytes) them (Strobel et al., 2004). The majority of drugs currently used in the clinic were derived from plants as either the purified natural product or synthetic analogs thereof (Grotewold, 2005; de Kok et al., 2008). For example, paclitaxel (Taxol) is a chemical originally isolated from the yew tree that is used for cancer therapy; it stabilizes microtubules and thereby inhibits cell division (Adams et al., 1993). Of interest, it has recently been shown that paclitaxel is not produced by cells of the yew tree but by fungi that inhabit the tree (Miller et al., 2008). We have found that although high doses of paclitaxel are toxic, low doses can have beneficial effects on neurons, effectively protecting them against degeneration in experimental models relevant to stroke and Alzheimer's disease (Furukawa and Mattson, 1995). One reason that many phytochemicals exert beneficial effects on cells is that they activate hormetic signaling pathways. Indeed, there is good evidence that metabolic pathways that generate hormetic phytochemicals evolved as a defense mechanism in which novel toxic chemicals are produced to protect the plant against insects and other organisms [see chapter The Fundamental Role of Hormesis in Evolution and Mattson and Cheng (2006)].

Several phytochemicals that can be classified as "botanical pesticides" have been shown to have interesting actions on the plasticity of brain cells in mammals. For example, curcumin (from the turmeric plant) enhances neurogenesis (the production of new nerve cells from stem cells) in rodents (Kang et al., 2006; Kim et al., 2008) and can improve learning and memory in mouse models of Alzheimer's disease (Frautschy et al., 2001). The flavanol epicatechin, which is present in high amounts in cocoa, enhanced retention of spatial memory and increased angiogenesis in the brains of mice (van Praag et al., 2007). Administration of sulforaphane attenuated the degeneration of photoreceptors in a mouse model of retinal degeneration (Kong et al., 2007). Resveratrol, a phytochemical present in red grapes, attenuated memory impairment and reduced neuronal degeneration in a mouse model of Alzheimer's disease (Kim et al., 2007). As described in the chapter The Devil Is in the Dose: Complexity of Receptor Systems and Responses, many of the hormetic effects of low doses of phytochemicals are mediated by their activation of specific adaptive stress response pathways involving transcription factors such as Nrf2, FOXO, and NF-κB. We believe that in the future, research efforts that use hormetic principles to identify, characterize, and move forward drugs to the clinic will greatly expand the pharmacopeia and so improve health throughout the lifespan.

Drugs

Drugs have historically been directed toward treating diseases and increasing performance, such as improving cognition, strengthening bones, growing hair, and accelerating healing times, among others, or in killing threatening organisms, such as harmful bacteria, fungi, yeasts, or viruses, or killing tumor cells. In the case of increasing performance, such responses are seen to occur at the lower end of the dose-response relationship. In the case of the killing activities, these are typically observed at the higher end of the dose-response relationship. Thus, the pharmaceutical industry is interested in the entire dose-response continuum, depending on the endpoints of concern.

The hormetic dose response is unique in that it indicates that there is biological activity below pharmacological and toxicological thresholds. The key concept to understanding this low-dose stimulation within a pharmaceutical framework is that it relates to biological performance. That is, the low-dose stimulation is a manifestation of gain the system permits and it is this function that many pharmaceuticals are designed to exploit. However, as will be shown, the capacity of gains in performance is limited to the percentage range rather than the fold range, and this is a consistent feature seen across all biological systems. For example, at low doses, aspirin inhibits platelet aggregation and clot formation, at moderate doses it has anti-inflammatory actions, and at high doses it can cause severe bleeding and even death.

In the case of hormetic dose responses, an assessment of many thousands of examples reveals that the maximum stimulation (i.e., biological performance) is typically modest, being only 30% to 60% greater than control values. This has been reported for a wide range of plants (Calabrese and Blain, 2009), microorganisms (Calabrese and Blain, 2005), cell lines (noncancer and cancer) (Calabrese, 2005), physiological functions, and responses of whole organisms. This prevalent observation is critical because it suggests strongly that pharmaceutical agents that act hormetically will have the magnitude of the response set within a maximum range of 30% to 60%, being constrained by inherent biological limits that we regard as a quantitative description of biological plasticity. Hormesis therefore sets limits on what drug and pharmaceutical companies can expect to achieve with drug treatments. This important limitation has not been widely recognized or discussed within the pharmacological and/or pharmaceutical literature. Yet it is the prime controlling feature affecting drug performance. This is a critical perspective, especially in terms of what the health care industry can expect to achieve and what patient expectations of treatment should be.

The issue of whether and how it may be possible to avoid the constraints of biological plasticity within the context of improving performance is an important theoretical question and one that is likely to be explored as the biological engineering of molecular systems evolves. However, it is important to recognize that these apparent biological constraints are very generalizable and long conserved within an evolutionary context. It is unknown what the biological implications would be if one were able to engineer around the constraints imposed by biological plasticity to achieve increases in biological performance that far exceeded the limits seen within the hormetic dose-response context. It would be expected that most systems will have redundant controls that could be activated should one be eliminated via a genetic or chemically induced manipulation. Based on this assumption, the prospect of chemically exceeding constraints imposed by biological plasticity is not likely to be easily achieved, but it nonetheless would be an intellectually compelling area of research.

Implications of the Quantitative Features of the Hormetic Dose Response for the Pharmaceutical and Nutraceutical Worlds

The quantitative features of the hormetic dose response can have important implications for the drug and nutraceutical industries. In fact, these quantitative features of the hormetic dose response are basic to this issue of "getting the dose right" because they guide scientists and physicians on the magnitude of the response, the range over which it works, and its relationship to the onset of toxicity.

Limiting the maximum increase in performance to only 30% to 60% has important implications for the selection of the biological model, study design, sample size, and statistical power, as well the number and spacing of doses within the hormetic zone. It affects how much a patient can be helped by the drug treatment that exploits the hormetic zone for a beneficial outcome. It also affects how much a patient may be damaged by a treatment within the low-dose zone that leads to undesirable effects. In the following we discuss how the concept of hormesis may affect a number of critical aspects of the process of hazard assessment.

Biological Model Selection

Hormesis can be very difficult to discern in experimental studies. One reason is that the magnitude of the hormetic stimulation is generally quite modest. If the biological model selected for study has high control-group variability for the endpoint of interest, then it creates a serious limitation with respect to being able to distinguish a hormetic response from chance or normal variability. Biological models with high background variability place considerable pressure on the need for larger sample sizes to have adequate statistical power. Failure to take this into account could lead to missing the potential to discern a statistically significant response. This is a common failure of many studies in which the investigator is unaware of the hormesis concept. In effect, the lack of a rigorous study design and/or statistical sensitivity prevents an evaluation of a potential low-dose hormetic response.

Another issue with biological models and hormesis is that drug treatments that are testing whether an agent can reduce disease incidence need to employ a control group with a relatively high disease incidence to test whether the drug treatment can reduce the disease incidence in a hormetic fashion. In the drug testing domain this is typically accomplished by inducing a disease and then assessing whether the treatment reduces the occurrence of the disease condition. This is perhaps the most cost-effective approach for addressing this issue, although it has its limitations. The induced disease may not be an excellent model of the human condition because it is artificially induced rather than occurring via normal developmental or aging processes. On the other hand, it may not be realistic to wait until older animals have developed the pathological condition of interest via normal aging processes due to the constraints imposed by time and other resources. The background disease concern is also important for the safety assessment of drugs and chemicals. Animal models with normally very low background disease incidence have typically been accepted for use for practical reasons, that is, it enhances the capacity for greater statistical power with fewer animals than if there were a higher background disease incidence. In these cases it is essentially impossible to assess hormetic hypotheses. Thus, the selection of the animal model, its normal disease incidence, and its variability are critical factors in assessing the dose-response and hormetic hypotheses. It is likely that if the hormetic concept had been operating in the 1960s and 1970s when current dose risk/risk assessment models were being selected, it could have altered the decisions on which particular mouse and rat strains were selected for study.

Drug-Testing Strategies

The hormetic dose response also has implications for the need to have an accurate estimation of the dose-response threshold because doses immediately lower than the threshold are where the key performance responses would be expected. Thus, careful preliminary experiments need to be conducted that reliably define the pharmacological/toxicological threshold zone. This would permit subsequent experiments to assess the possibility of hormetic dose responses in an efficient manner. This would require additional time and resources, which would have to be balanced off against the value of data where the expected hormetic performance enhancement would be predicted to occur. Focusing drug development on a specific molecular target (receptor, enzyme, etc.) rather than the relevant biological endpoint (cardiac function, cognition, inflammatory process, etc.) may result in failure to identify drugs that exert hormetic effects at doses other than the dose that affects the molecular target. Thus, a drug may bind to a target of interest but exert hormetic effects via an unrelated mechanism of action.

Once a reliable estimate of the threshold response is known, then it may be possible to design follow-up studies that can adequately assess the hormetic stimulatory zone. The investigator would usually desire to have several (at least four) doses within the hormetic response zone. This will provide information on the maximum stimulation response but also on the width of the stimulatory response zone. This information may have not only statistical implications, but also clinical implications. For example, a very broad stimulatory range would provide the physician with a wider drug concentration target and may be useful in avoiding concerns with interindividual variability in response to drug-induced toxicities.

Drug Potency

The hormetic concept is also relevant to the issue of drug potency. The quantitative features of the drug response will be similar regardless of drug potency. Even if an agent is many times more potent than another, the principal difference is that the response of the more potent agent will simply be shifted further to the left on the dose-response graph than that of the less potent agent, but the quantitative features of the hormetic response would be similar for both chemicals. This concept is important because the desired increase in performance (i.e., benefit) would be independent of drug potency.

Drug–Drug Interactions

The magnitude of the increase in biological performance (e.g., increases in cognition, hair growth, bone strengthening) is limited by the constraints imposed by biological plasticity, which is quantified by the hormetic dose response. This is the case even when multiple drugs are used with the intention of achieving either an additive or a synergistic response. This is another concept that has emerged from an assessment of hormetic dose responses. There in fact can be additive and synergistic responses, but the outcome response will be constrained by the so-called 30% to 60% "rule," keeping the response to the modest hormetic level. In practical terms the drug–drug interaction is best observed at the level of drug dose rather than at the effect level that is typically seen within a toxicological setting.

These observations may have important implications in a clinical framework. For example if a drug improved cognition by about 30% in a patient, it is unlikely that a drug combination would increase this response. A drug combination treatment may, in fact, achieve this percentage increase by using a far lower dose, thereby acting synergistically. Its advantage is that there would be the likelihood of fewer possible side effects because of the lower drug doses used, but it would not be expected to result in an increased cognition performance over the optimal treatment of a single well-performing drug whose response approaches the limits of biological plasticity.

Clinical Trials and Hormesis

The use of a heterogeneous group in a clinical trial presents serious concerns to investigators. Because there may be considerable interindividual variation in response to the drug, it is expected that the subjects will have a response that is distributed across the entire hormetic zone and possibly into the beginnings of toxicity. For those subjects in the toxicity zone the exposure is unfortunately too high. When the response data are analyzed, the entire grouping of subjects may be averaged. This can result in a confusing picture to the investigators and governmental regulators. The clinical trial therefore needs to take the concept of hormesis into account because even those at higher risk will display the low-dose stimulation but at a dose lower than the average subject. However, because clinical trials usually use only one or two doses, this can contribute to missing possible beneficial effects and drug failure.

Although the concept of hormesis and its requirements for larger numbers of doses and more rigorous study designs may not be particularly welcome to a pharmaceutical industry that aims at reducing drug development costs, they may be balanced against the savings of obtaining fewer data points and having a compound fail at early or even more costly stages of development because of an incorrect assumption that the dose response was linear or sigmoidal. Consequently, it is possible that many drug discovery and developmental activities have been stopped or abandoned principally because the optimal concentrations or doses were not tested and not because the target was wrong or the agent was ineffective or toxic. This will occur more frequently as attempts are made to reduce costs in clinical trials by testing only one or two doses. In contrast, the hormetic perspective would employ the early stages of research and development up to Phase II to clarify the dose response employing a broad range of doses and frequent intervals to better define the relationship between drug doses to clinical benefit and adverse or toxic events. It should be noted that one in five drugs approved by the FDA between 1980 and 1999 had dose adjustments postapproval; 20% of these had the dose increased, while 80% had the dose decreased, thereby supporting the need to get the dose right (Cross et al., 2002).

Potentially Harmful Hormetic Responses

Although the foregoing discussion focused on exploiting the hormetic dose response to enhance biological performance, the hormetic dose response can also have harmful medical implications in the case of antibiotics and antitumor drugs, in which case it may enhance the "performance" of the harmful microbes or tumor cells to proliferate. Although the general clinical plan in these instances is to maintain a sufficiently high drug titer to kill the threatening organisms/cells, it is possible and even likely that drug concentrations will decrease, entering into the hormetic zone. A recent series of studies with several thousand possible antitumor agents revealed that essentially all the drugs passing the a priori entry criteria induced responses consistent with the hormetic dose-response model (Calabrese et al., 2008). Thus, the hormetic concept is a key feature whether one is interested in achieving the beneficial or preventing the harmful effects of enhancing biological performance.

The ideal antitumor drug would be one that combined two features— cell killing and not inducing a cell proliferative response at lower concentrations, that is, the hormetic dose response. No published studies exist on the structural determinants of hormetic responses for antitumor agents. The use of antitumor agents with a relatively short biological half-life would reduce the potential for agent-induced hormetic effects to have an adverse effect. This was recognized in the case of the antitumor agent suramin, which clearly induces hormetic biphasic dose responses in various human tumor cells while having a rather prolonged residence time of about 40 to 50 days within the human body (Foekens et al., 1992). Such a prolonged period in the human body could result in tumor cells being exposed to this drug while the concentration was within the hormetic zone. Thus, in the case of the antitumor drugs, failure to recognize the potential for a hormetic dose response may be a contributory factor in tumor resurgence and treatment failure. The treatment checklist should no longer relate only to the domain of tumor-killing potential and likely side effects but also should include the potential for hormetic dose responses within the context of a drug's biological half-life. However, whether hormetic effects at low doses of antitumor agents actually contribute to the survival risks of cancer patients is a hypothesis at present. There have been no detailed studies published on the modeling of such effects in animal models or humans or experimental studies that have directly addressed this point.

Despite the lack of data with respect to the role of hormetically acting anticancer drugs on patients, this concept has been strongly supported within the area of antibiotics starting in the 1940s. In these instances, investigators demonstrated that low doses of antibiotics could act hormetically, increasing the proliferation of bacteria at low concentrations while inhibiting their growth at higher concentrations in cell culture studies. However, in vivo follow-up studies indicated that very low doses of common antibiotics such as penicillin and streptomycin could enhance microbe-induced mortality in mice, whereas at higher doses the antibiotics prevented mortality (Randall et al., 1947; Welch et al., 1946). These observations support the hypothesis that low doses of antibiotics and antitumor agents need to be assessed for their capacity to induce hormetic dose responses and to further assess such behavior within an in vivo framework. As in the case with antitumor agents, the hormetic implications of antibiotics have also been ignored.

The potential for adverse effects of hormetic low-dose stimulation can be extended well beyond those cells that medical treatments were designed to inhibit or kill, as discussed earlier. Hormetic biphasic dose-response relationships have been implications in conditions such as benign prostate hyperplasia (BHP), detached retina, and Dupuytren's contracture (Murrell et al., 1990, 1989). In each of these conditions there is a modest overstimulation of cellular proliferation, which can lead to the recognized pathology. In each case the cell proliferation response is biphasic, with a low-dose stimulation and a high-dose stimulation.

Cardiac glycosides are known to enhance the proliferation of a wide range of cell types in a biphasic dose-response manner consistent with the quantitative features of the hormetic dose-response model. Although this was first reported with immune cells, it was subsequently seen with smooth muscle for the human umbilical cord (Abramowitz et al., 2003), for the canine saphageous vein (Aydemir-Koksoy et al., 2001), and for a rat cell line (A7r5), as well with rat epithelial cells (Dmitrieva and Doris, 2003) and HeLa cells (Ramirez-Ortega et al., 2006). It was also well known that the cardiac glycosides biphasically affected the sodium pump, that is, stimulating activity at low concentrations and decreasing activity at higher concentrations (Godfraind and Ghysel-Burton, 1977). In general, these studies were focused on clarifying how the sodium pump functioned and its capacity to act in ways that may not employ the well-established ionic shift, thereby implying a multiplicity of functions for the sodium pump, such as diverse message signaling, with wide-ranging clinical implications.

The fact that cardiac glycosides were particularly active in affecting the proliferation of smooth muscles cells from the aforementioned tissues in a biphasic manner suggested that this response may be extended to smooth muscles of the prostate. Concentrations in the therapeutic zone of the cardiac glycoside ouabain increased cell proliferation of smooth muscle of the prostate gland of individuals with BHP (Chueh et al., 2001). Such findings suggested that these modest increases may have clinical implications, given that smooth muscle makes up nearly 40% of the area density of BPH tissues. In fact, a small increase in prostate size may significantly affect clinical symptoms in patients with BPH. Likewise, a modest reduction in prostate size (as measured by volume) of 30% has been consistently shown to improve clinical symptoms in a significant manner (Chueh et al., 2001). Consequently, the hormetic biphasic dose response may have an important role in affecting the occurrence of BHP in such exposure situations.

In the case of intraocular proliferative diseases such as proliferative vitreopathy (PVR), progressive traumatic traction retinal detachment (PTTRD), and intraocular neovascularization there is excessive accumulation of fibrous vascular tissue within the eye. What these conditions have in common is the rapid and uncontrolled proliferation of nonneoplastic cells within or about the eye. The induced damage results from fibrocellular proliferation, active contraction of cellular membranes, and the formation of cross-lineages of newly formed collagen by fibroblasts and myofibroblasts. These observations led to the suggestion that nontoxic pharmacological agents that inhibit the growth of rapidly proliferating cells may have therapeutic utility in the treatment of such conditions. Based on this reasoning, numerous studies were undertaken to find drugs that could inhibit the proliferation response while not being harmful to ocular tissue. Consideration was given to cancer chemotherapeutic agents, such as 5FU (fluorouracil) and daunomycin, because they displayed potent antiproliferative effects. The use of corticosteroids was also proposed because they too had the capacity to inhibit mitosis (Blumenkranz et al., 1984) yet were far less toxic than the chemotherapeutic antitumor drugs. Similarly, a wide range of steroids, nonsteroidal anti-inflammatory drugs, and antimetabolites and biopeptides have been tested. An assessment of these investigations indicated that hormetic-like biphasic dose responses commonly occurred independent of the chemical agent and biological model system used. The quantitative features of the dose response were typical of hormetic dose responses, with the maximum stimulatory response being only about 30% to 60% greater than the control. However, there was the potential for a highly variable stimulatory range, an observation occasionally seen in other experimental systems. These findings illustrate the need to incorporate a broadened dose-response perspective when addressing complex clinical problems relating to diseases associated with cell proliferation. In the past it may have been important to only be concerned with whether the drug could inhibit cell proliferation while not being toxic to ocular tissues. It is now necessary to add a third factor, that is, the potential for hormetic responses and the quantitative features of the dose responses, in differentiating among candidate therapeutic agents.

Hormesis in the Pharmaceutical Industry

In 2008 Calabrese (2008c) published a detailed summary of the role of hormesis in medicine. This paper revealed that hormetic effects underlie the responses of many drug classes affecting various human clinical conditions. This section provides an overview of selected hormetic medical conditions.

Anxiolytic Drugs

A major area of pharmaceutical research is in the area of anxiolytic drugs, that is, drugs that treat patients with various anxiety disorders. Preclinical studies in the area of anxiolytic drugs typically involve experiments with rodents. The object of these studies is to find drugs that are effective in making the animals do what they are normally afraid to do. More specifically, a rodent strongly favors staying in dark areas as compared to lighted areas. If a drug could motivate the rodent to spend more time in the lighted areas, it would be interpreted as reducing its anxiety. There are many ways for the pharmacologist to test the anxiolytic effects of drugs in rodents (for a review see Calabrese, 2008d). The investigator determines how much time the rodent spends in the lighted area or the number of entries into the lighted zone the rodent makes as compared to the dark zone. If the drug gets the rodent to do what it normally would not do, then this is used as a measure of its anxiolytic properties and can be quantitatively assessed. When such drugs are tested over a broad dose range, the typical response is biphasic, with more time being spent in the lighted areas at lower doses, whereas the reverse is the case with higher doses. This is a general pattern of response that is fully consistent with the hormetic dose-response model. The hormetic biphasic dose response occurs independent of animal model, chemical used, and molecular pathway by which the agent acts. These observations indicate that the hormetic dose response is central to the field of anxiolytic drugs. As suggested earlier, the quantitative features of the hormetic dose response for anxiolytic drugs are consistent with hormetic responses in other biological systems.

One of the important findings is that the reduction in anxiety is the hormetic stimulatory response. It is the measure of biological performance discussed earlier. Such anxiolytic drugs at low dose can increase the amount of time spent in the lighted zones by about 30% to 60% at maximum (Calabrese, 2008d). When drugs pass the preclinical testing phase with rodents, the target dose area is within the hormetic zone. Thus, the concept of hormesis actually drives the area of anxiolytic drug discovery, development, and clinical application. Despite the fact that the hormetic dose response is commonly observed in rodent studies of anxiolytic effects, it was not until recently that this association was brought forward (Calabrese, 2008d). Thus, researchers in this area over the last half century had repeatedly observed the biphasic nature of the anxiolytic drug response but failed to see it as part of a general biological response pattern of low-dose effects.

Antiseizure Drugs

Antiseizure drugs are studied using a variety of animal models within the preclinical setting. Numerous drugs are also used to induce different types of seizure conditions, depending on the interest of the investigator. These experimental settings are therefore designed to reliably induce the type of seizure desired in the animal model at a consistent level of dose. The goal of preclinical testing is to determine whether candidate antiseizure drugs can make it more difficult to induce the seizure response in the standard protocol. This would be observed if the amount of drug inducing the seizures had to be increased to achieve the standard inducing effect. That is, the threshold dose would have to be increased. The greater the increase in the threshold dose, the greater is the potential for the agent to act as an antiseizure drug. An evaluation of such preclinical testing reveals that drugs that can induce an increase in the seizure threshold do so in a manner that is fully consistent with the hormetic dose response. That is, these drugs increase the threshold of the inducing agent by about 30% to 60% at maximum, whereas at higher doses such antiseizure agents become proseizure agents, actually lowering the threshold in a dose-dependent manner (Calabrese, 2008e). The overall response clearly follows the biphasic dose response typical of hormetic responses for other biological models, chemicals, and endpoints.

Male Sexual Dysfunction

Another area exhibiting hormetic dose responses is that of drugs that address the issue of male sexual dysfunction. As with other drugs intended for human consumption, the process involves considerable testing with animal models. For the current generation of erectile dysfunction drugs, male rats and male beagle dogs were commonly employed in the early testing phases (Calabrese, 2008c). There are numerous chemical classes that have been experimentally shown to enhance male sexual performance (e.g., penile erection, ejaculatory functions), generally showing a hormetic-like biphasic dose response. These agents may affect a range of receptor systems, with particular attention having been given to dopamine, α-adrenergic agents, serotonin, opioids, nitric oxide, cholinergic agents, histamine, prostaglandin, oxytocin, and others. These data indicate that such agents enhance sexual performance in a variety of ways, as measured by stiffness of the penis, duration of the erection, ability to penetrate, and amount of semen ejaculated, among other endpoints of interest. Although most papers with evidence of a biphasic dose-response relationship for a male sexual behavior endpoint have acknowledged and discussed their occurrence, few have offered hypotheses that might account for these general and consistent responses.

Diabetes

It is well established that a modest amount of exercise is associated with a reduced occurrence of type 2 diabetes. A number of studies have shown that modest exercise generates low levels of oxidant stressor compounds that act to enhance the capacity of insulin to affect the uptake of glucose by a variety of cells. These findings suggest that a low degree of oxidant stress may be necessary to enhance insulin functioning, making this another example of hormesis. These findings also suggest that efforts to reduce oxidant radicals by the ingestion of large amounts of antioxidant vitamins may prevent the exercise-induced benefit. The flip side of the coin for hormesis is that higher levels of oxidant stress may reverse the process and make the condition worse. Although this remains to be demonstrated, it would not be unexpected, given that the hormetic dose response clearly indicates that for such performance endpoints there is a dose optimum, and extremes on either side of this response optimum are likely to lessen performance considerably.

Memory/Cognition

Most drugs that increase memory in animal models do so via a U-shaped dose-response relationship. This was first observed in the 1960s by McGaugh and Petrinovich (1965) in studies with both bright and dull rodents. In their study they employed the drug physostigmine, an anticholinesterase, based on the hypothesis that this drug treatment would result in an increase in acetylcholine within the synaptic zone and facilitate the learning process. Their findings revealed that both the bright and dull mice had their performance increased at lower doses, whereas performance declined as the dose increased. Although some were initially skeptical that an exogenous chemical could increase memory performance, this seminal investigation opened the floodgates of research into the chemical manipulation of memory, eventually leading the development and marketing of such drugs for cognitively impaired patients, such as those with Alzheimer's disease (Calabrese, 2008f). The U-shaped dose-response relationship is widely reported for drugs that enhance memory, regardless of the chemical class of the effective agents. Each of the six drugs approved by the FDA for the treatment of Alzheimer's disease displays the standard type of hormetic biphasic dose-response relationship.

Osteoporosis

Loss of bone density leading to osteoporosis is a serious public health issue for postmenopausal women who have had children. About 10 million women older than 50 years of age in the United States alone have osteoporosis, with nearly twice that number being at risk as a result of low bone mass (Foreman, 2001). Numerous pharmacologically based strategies have been employed to prevent the onset of osteoporosis. Of particular interest are the biphosphonates, which are widely used. These agents prevent bone resorption and therefore are used to treat osteoporosis. Although initial studies emphasized the capacity of bisphosphonates to have direct inhibitory effects on mature osteoclasts, Giuliani et al. (1998) determined that they also have a direct effect on osteoblasts, providing an alternative target for bisphosphonate-induced beneficial effects on the bone formation process. Using the bisphosphonate alendronate, Giuliani et al. reported that it displayed a biphasic dose-response relationship with the stimulation of CFU-F (colony forming units-fibroblasts) occurring at lower than 100 nM. The maximum stimulatory response was nearly 80% greater than that of the controls. Similar findings were also reported for other bisphosphonates. Mechanistically oriented studies now suggest that biphosphonates affect an increase in the number of mesenchymal bone marrow cells committed to the osteoblast phenotype. The general effectiveness of these agents may represent a beneficial influence of osteoblast precursors to complement the previous recognized high-dose inhibitory effect on osteoblastic bone resorption.

Hormesis and Neutraceuticals

Dietary supplements ranging from vitamins and minerals to herbal preparations and purified and synthetic phytochemicals are now widely used by many health-conscious individuals in modern societies (Triggiani et al., 2006). Accordingly, a large "neutraceutical" industry has evolved to develop and market such natural chemicals. There is a large body of literature demonstrating beneficial effects of many different phytochemicals in animal models of a range of diseases (cancers, cardiovascular and cerebrovascular diseases, inflammatory conditions, neurological disorders, etc.). However, despite the good evidence that consumption of vegetables and fruits reduces disease risk, the data from human studies is less convincing regarding the health benefits of individual phytochemicals. One possible explanation for differences in outcomes of studies in animals and humans is that the magnitude of hormetic effects is typically in the range of 20% to 50% (see the chapter Hormesis: Once Marginalized, Evidence Now Supports Hormesis as the Most Fundamental Dose Response), and therefore it is difficult to discern significant effects of hormetic agents when there are high levels of variability of genetic and environmental factors among individuals in the test population. Thus, there is low interindividual variability in laboratory animals (e.g., inbred lines of mice maintained in identical housing environments) and high variability among humans, who come from different genetic backgrounds and living environments.

We believe that the development of large-scale biological screens for natural products that activate specific adaptive cellular stress response pathways will result in the identification of novel chemicals with prophylactic and therapeutic efficacy in humans (Fig. 1). Modern technologies in molecular and cellular biology provide the opportunity to develop cell-based screens to identify hormetic chemicals. For example, cell lines (normal or cancerous) can be transfected with plasmids containing transcription factor–specific reporter constructs in which transactivation by the endogenous transcription factor results in the expression of a reporter protein such as luciferase or green fluorescent protein. Transcription factors that are known to mediate hormetic cellular responses (NF-κB, Nrf2, FOXO's, CREB) are good targets for screening extracts from plants or microorganisms. Other screens could involve measuring levels of cytoprotective proteins expressed in response to activation of hormetic pathways; these might include antiapoptotic Bcl-2 or IAP proteins, protein chaperones such as heat-shock protein 70 (HSP70), ARE-induced proteins

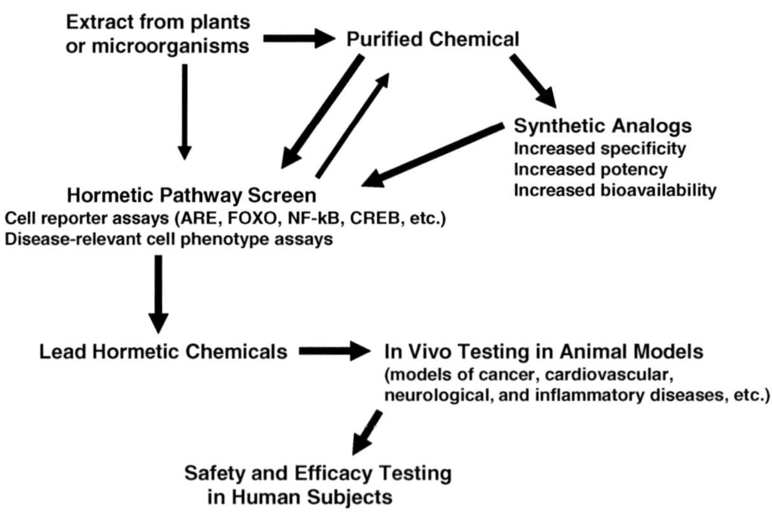

Fig. 1 Strategy for the identification and preclinical and clinical testing of hormetic chemicals. ARE, antioxidant response element; CREB, cyclic AMP response element–binding protein; FOXO, member of the forkhead transcription factor family; NF-κB, nuclear factor κB

such as HO-1 and NQO1, and antioxidant enzymes such as Mn-superoxide dismutase and glutathione peroxidase. Chemicals that activate such pathways at subtoxic concentrations can then be moved forward to in vivo models of diseases. In addition, analogs of the lead chemicals can be synthesized and tested with the goal of identifying agents with improved potency, selectivity, and biological availability (Fig. 1). Finally, lead chemicals would be tested for safety and treatment efficacy in human subjects.

Hormetic Mimetics

An emerging area of hormetic research is that of molecular mimetics—agents than can activate hormetic pathways and reverse to some degree functional deficits of age-related disease processes that have common molecular foundations in altered gene expression and protein structure (Smith-Sonneborn, 2008). Numerous chemical mimetics of the beneficial effects of hormetic responses due to exposures from heat, radiation, caloric restriction, cold, and hunger have been shown to be inducers of stress resistance. It is important to note that this has been achieved without the exposure of the various biological systems to the environmental stressor agent itself. This represents a conceptual transformation of the hormesis concept and one that creates an expansive array of pharmaceutical opportunities. Smith-Sonneborn (2008) reported the occurrence of hormetic mimetics from a wide range of chemical groupings, involving small nucleotide SOS signals, dipeptides, thiols, metals, ethanol, and conserved peptide sequences from various models that regulate cytokines, cellular immunity, and central and peripheral neural regulatory pathways affecting numerous critical endpoints, such as blood pressure maintenance, heart rate, metabolic pathway regulation, and age-related disease susceptibility, by activating various protein and DNA functions.

How a hormetic mimetic may work can be seen in the case of ethanol. Studies have shown that a moderate ethanol pretreatment effects a severalfold increase in brain levels of HSP70 and can prevent β-amyloid peptide–induced neurotoxicity and apoptosis in hippocampal-entorhinal slice cultures. These findings suggest the existence of molecular mechanisms that may account for the protective effects of moderate ethanol consumption against Alzheimer's disease dementia (Belmadami et al., 2004). Similar protective effects of ethanol have been reported in behavioral deficit studies induced by cerebral ischemia/reperfusion while also protecting against neuronal cell death and dendrite degeneration.

Other types of hormetic mimetics that affect some of the beneficial effects of caloric restriction include glycolytic inhibitors, lipid regulatory agents, antioxidants, sirtuin regulators, autophagic enhancers, insulin-specific gene modulators, and weight-loss drugs (Smith-Sonneborn, 2008). Other hormetic mimetics may include agents that prevent glycation that leads to altered protein cross-linkages. Such cross-linkages have been associated with numerous diseases, such as diabetes, heart disease, and neurodegenerative clinical conditions. It is hoped that hormetic mimetics may intervene before, during, or after the formation of glycation products, thereby preventing the onset and/or progression of various chronic diseases (Smith-Sonneborn, 2008).

Another area of hormetic mimetics involves agents that activate hibernation-like processes that typically reduce stress from depleted energy stores, intracellular acidosis, hypoxia, hypothermia, cell volume shift, and muscle wasting. A hibernation induction trigger (HIT), an opioid-like peptide, has been reported in hibernating woodchucks. This agent protects against skeletal muscle damage resulting from hypoxia/reperfusion. Several HIT mimetic candidates have induced a broad range of adaptive responses to different types of stress-induced conditions and/or toxic agents in a variety of animal models and cell types, displaying multisystem (e.g., cardiovascular, neurological) protective effects, as well as adaptive behavioral responses (Smith-Sonneborn, 2008). Such agents also display these protective effects in both pre-and postconditioning hormetic exposure protocols. This suggests that the hormetic mimetics may have application in preventive as well as therapeutic settings.

Because all systems are likely to be affected by hormetic mimetics, the potential range of applications is extraordinary. It can be expected that hormetic mimetics will be of potential utility in emergency trauma as would occur in wartime, accidents, and critical surgical procedures, especially in relation to concerns associated with ischemic stress (Smith-Sonneborn, 2008).

Discussion

The hormetic pharmacy represents the recognition that the concept of hormesis plays a significant, if not dominant, role in the process of drug discovery, development, clinical evaluation, and therapeutic application. This and other chapters of this book have shown that the concept of hormesis is critical for a broad spectrum of drugs that affect all systems of the body. Moreover, it is the low-dose stimulatory response that pharmaceutical companies target for essentially all of their drugs that deal with improving biological performance. Although this has been demonstrated with comprehensive documentation in numerous cases, the striking fact is that the pharmaceutical industry and its underlying academic, governmental, and industry research base has generally failed to see the broadly integrative and conceptual perspective that has emerged on the nature of the dose response. Rather, their focus has been far more restricted, being directed to specific drugs, their successes, failures, mechanisms, side effects and economic implications. Although this is a critical perspective, their overarching research myopia is a dominant factor contributing to their failure to recognize the basic perspectives offered by the hormetic dose-response model and its capacity to guide and enhance the discovery, development, and approval of candidate drugs.

This chapter has documented the fact that the hormesis concept offers a quantitative measure of the limits of biological plasticity of all organisms and at all of levels of biological organization. The constraints of plasticity then determine the quantitative features of the drug response for all performance-related endpoints, thereby controlling and defining the magnitude of drug -induced responses. Thus, it is the hormesis concept that sets the biological rules of the pharmaceutical "game," dictating the expected benefits and challenges that the industry faces in demonstrating such benefits within the context of preclinical animal studies, as well as the clinical trial. The quantitative features of the hormetic dose response are highly predictable, not only with respect to the magnitude of response, but also for the width of the stimulatory response and the relationship of the enhancement of performance to the onset on possible toxicity on the dose-response continuum. This is a powerful biological trifecta for the guidance of the pharmaceutical industry in theory as well as practice.

The hormetic pharmacy will also enlighten physicians and patients about the possibility of hormetically induced adverse effects for the spectrum of drugs that enhance the proliferation of harmful microbes, viruses, tumor cells, and other tissues, which could lead to various pathologies. In this respect, the hormetic pharmacy will help to educate individuals about the need to faithfully follow treatment recommendations to maintain the drug dose in the optimal zone rather than have it work against the patient. The hormesis concept should therefore help in improving the overall chances of treatment success and not only diminish the risk of treatment failure but also prevent new, unexpected, and harmful side effects.

The hormetic pharmacy not only represents the present, but also points to the future of the field and its growing impact on health care, both institutionally and on the personal level. As discussed earlier, it is expected that a new field, that of hormetic mimetics, will be created and will transform the pharmaceutical and health care industries in the 21st century. Such hormetic mimetics will be able to activate adaptive pathways and prevent and reverse chemically induced injuries and harmful effects from a variety of traumas, including heart attacks, strokes, brain injuries, and other devastating human injures and diseases. The emerging field of hormetic mimetics will also transform the field of geriatric medicine, in that it will be particularly focused on ameliorating the effects of normal and accelerated aging patterns. Although this perspective speaks to a future decades from the present, considerable research is ongoing on the identification and testing of various types of hormetic mimetics, offering the promise of significantly improved functioning with a profoundly minimized risk of undesirable side effects.

Acknowledgments This work was supported by the Intramural Research Program of the National Institute on Aging, National Institutes of Health.

References

Abramowitz J, Dai C, Hirschi KK, Dmitrieva RI, Doris PA, Liu L, Allen JC (2003). Ouabain- and marinobufagenin-induced proliferation of human umbilical vein smooth muscle cells and a rat vascular smooth muscle cell lines, A7r5. Circulation 108: 3048–3053

Adams JD, Flora KP, Goldspiel BR, Wilson JW, Arbuck SG, Finley R (1993) Taxol: a history of pharmaceutical development and current pharmaceutical concerns. J Natl Cancer Inst Monogr 15: 141–147

Arumugam TV, Gleichmann M, Tang SC, Mattson MP (2006) Hormesis/preconditioning mechanisms, the nervous system and aging. Ageing Res Rev 5: 165–178

Aydemir-Koksoy A, Ambramowitz J, Allen JC (2001) Ouabain-induced signaling and vascular smooth muscle cell proliferation. Biol Chem 276: 46605–46611

Bates RS (1965) Scientific societies in the United States, 3rd ed. Cambridge, MA: MIT Press.

Belmadani AS, Kumar S, Schipma M, Collins MA, Neafsey EJ (2004) Inhibition of amyloid-beta–induced neurotoxicity and apoptosis by moderate ethanol preconditioning. Neuroreport 15: 2093–2096

Blumenkranz MS, Clafin A, Hajek AS (1984). Selection of therapeutic agents for intraocular proliferative disease. Arch Ophthalmol 102: 598–604

Calabrese EJ (2005) Cancer biology and hormesis: human tumor cell lines commonly display hormetic (biphasic) dose responses. Crit Rev Toxicol 35: 463–582

Calabrese EJ (2008a) Converging concepts: adaptive response, preconditioning, and the Yerkes–Dodson law are manifestations of hormesis. Aging Res Rev 7: 8–20

Calabrese EJ (2008b). Hormesis: why it is important to toxicology and toxicologists. Environ Toxicol Chem 27: 1451–1474.

Calabrese EJ (2008c) Hormesis and medicine. Br J Clin Pharmacol 66: 594–617

Calabrese EJ (2008d) An assessment of anxiolytic drug screening tests: hormetic dose responses predominate. Crit Rev Toxicol 38: 489–542

Calabrese EJ (2008e) Modulation of the epileptic seizure threshold: implications of biphasic dose responses. Crit Rev Toxicol 38: 543–556

Calabrese EJ (2008f) Alzheimer's disease drugs: an application of the hormetic dose-response model. Crit Rev Toxicol 38: 419–452

Calabrese EJ, Blain R (2005). The occurrence of hormetic dose responses in the toxicological literature, the hormesis database: an overview. Toxicol Appl Pharmacol 202: 289–301

Calabrese EJ, Blain, RB (2009). Hormesis and plant biology. Environ Pollut 157: 42–48

Calabrese EJ, Stanek EJ III, Nascarella MA, Hoffmann GR (2008). Hormesis predicts low-dose responses better than threshold models. Int J Toxicol 27: 369–378

Carter CS, Hofer T, Seo AY, Leeuwenburgh C (2007) Molecular mechanisms of life- and health-span extension: role of calorie restriction and exercise intervention. Appl Physiol Nutr Metab 32: 954–966

Chigurupati S, Son TG, Hyun DH, Lathia JD, Mughal MR, Savell J, Li SC, Nagaraju GP, Chan SL, Arumugam TV, Mattson MP (2008) Lifelong running reduces oxidative stress and degenerative changes in the testes of mice. J Endocrinol 199: 333–341

Chueh S-C, Guh J-H, Chen J, Lai M-K, Teng C-M (2001). Dual effects of ouabain on the regulation of proliferation and apoptosis in human prostatic smooth muscle cells. J Urol 166: 347–353

Colman RJ, Beasley TM, Allison DB, Weindruch R (2008) Attenuation of sarcopenia by dietary restriction in rhesus monkeys. J Gerontol A Biol Sci Med Sci 63: 556–559

Cross J, Lee H, Westelinck A, Nelson J, Grudzinskas C, Peck C (2002). Postmarketing drug dosage changes of 499 FDA-approved new molecular entities, 1980–1999. Pharmacoepidemiol Drug Saf 11: 439–446

de Kok TM, van Breda SG, Manson MM (2008) Mechanisms of combined action of different chemopreventive dietary compounds: a review. Eur J Nutr 47 (Suppl 2): 51–59

Dmitrieva R, Doris PA (2003). Ouabain is a potent promoter of growth and activator of ERK1/2 in ouabain-resistant rat renal epithelial cells. Biol Chem 278: 28160–28168

Drory Y, Kramer MR, Lev B (1991) Exertional sudden death in soldiers. Med Sci Sports Exerc 23: 147–151

Foekens JA, Sieuwerts AM, Stuurman-Smeets EMJ, Dorssers LCJ, Berns EMJJ, Klijn JGM (1992). Pleiotropic actions of suramin on the proliferation of human breast cancer cells in vitro. Int J Cancer 51: 439–444

Fontana L (2008) Calorie restriction and cardiometabolic health. Eur J Cardiovasc Prev Rehabil 15: 3–9

Foreman J (2001). A 'cure' for osteoporosis may be near. *Boston Globe*, April 24, 2001, pp. C1, C4.

Frautschy SA, Hu W, Kim P, Miller SA, Chu T, Harris-White ME, Cole GM (2001) Phenolic anti-inflammatory antioxidant reversal of Abeta-induced cognitive deficits and neuropathology. Neurobiol Aging 22: 993–1005

Furukawa K, Mattson MP (1995) Taxol stabilizes [Ca2+]i and protects hippocampal neurons against excitotoxicity. Brain Res 689: 141–146

Giuliani N, Pedrazzoni M, Negri G, Passeri G, Impicciatore M, Girasole G (1998) Bisphosphonates stimulate formation of osteoblast precursors and mineralized nodule in murine and human bone marrow cultures in vitro and promote early osteoblastogenesis in young and aged mice in vivo. Bone 22: 455–461

Godfraind T, Ghysel-Burton J (1977) Binding sites related to ouabain-induced stimulation or inhibition of the sodium pump. Nature 365: 165–166

Grotewold E (2005) Plant metabolic diversity: a regulatory perspective. Trends Plant Sci. 10: 57–62

Hubbard RW (1990) The role of exercise in the etiology of exertional heatstroke. Med Sci Sports Exerc 22: 2–5

Johnson JB, Summer W, Cutler RG, Martin B, Hyun DH, Dixit VD, Pearson M, Nassar M, Telljohann R, Maudsley S, Carlson O, John S, Laub DR, Mattson MP (2007) Alternate day calorie restriction improves clinical findings and reduces markers of oxidative stress and inflammation in overweight adults with moderate asthma. Free Radic Biol Med 42: 665–674

Kang SK, Cha SH, Jeon HG (2006) Curcumin-induced histone hypoacetylation enhances caspase-3-dependent glioma cell death and neurogenesis of neural progenitor cells. Stem Cells Dev 15: 165–174

Kim D, Nguyen MD, Dobbin MM, Fischer A, Sananbenesi F, Rodgers JT, Delalle I, Baur JA, Sui G, Armour SM, Puigserver P, Sinclair DA, Tsai LH (2007) SIRT1 deacetylase protects against neurodegeneration in models for Alzheimer's disease and amyotrophic lateral sclerosis. EMBO J 26: 3169–3179

Kim SJ, Son TG, Park HR, Park M, Kim MS, Kim HS, Chung HY, Mattson MP, Lee J (2008) Curcumin stimulates proliferation of embryonic neural progenitor cells and neurogenesis in the adult hippocampus. J Biol Chem 283: 14497–14505

Kong L, Tanito M, Huang Z, Li F, Zhou X, Zaharia A, Yodoi J, McGinnis JF, Cao W (2007) Delay of photoreceptor degeneration in tubby mouse by sulforaphane. J Neurochem 101: 1041–1052

Lee J, Duan W, Mattson MP (2002) Evidence that brain-derived neurotrophic factor is required for basal neurogenesis and mediates, in part, the enhancement of neurogenesis by dietary restriction in the hippocampus of adult mice. J Neurochem 82: 1367–1375

Liu D, Chan SL, de Souza-Pinto NC, Slevin JR, Wersto RP, Zhan M, Mustafa K, de Cabo R, Mattson MP (2006) Mitochondrial UCP4 mediates an adaptive shift in energy metabolism and increases the resistance of neurons to metabolic and oxidative stress. Neuromolecular Med 8: 389–414

Maswood N, Young J, Tilmont E, Zhang Z, Gash DM, Gerhardt GA, Grondin R, Roth GS, Mattison J, Lane MA, Carson RE, Cohen RM, Mouton PR, Quigley C, Mattson MP, Ingram DK (2004) Caloric restriction increases neurotrophic factor levels and attenuates neurochemical and behavioral deficits in a primate model of Parkinson's disease. Proc Natl Acad Sci USA 101: 18171–18176

Mattson MP, Cheng A (2006) Neurohormetic phytochemicals: low-dose toxins that induce adaptive neuronal stress responses. Trends Neurosci 29: 632–639

McGaugh JL, Petrinovich LF (1965). Effects of drugs on learning and memory. Int Rev Neurobiol 8: 139–196

Miller K, Neilan B, Sze DM (2008) Development of Taxol and other endophyte produced anti-cancer agents. Recent Patents Anticancer Drug Discov 3: 14–19

Murrell GAC, Francis MJO, Bromley L (1989). Oxygen free radicals stimulate fibroblast proliferation. Biochem Soc Trans 484

Murrell GAC, Francis MJO, Bromley L (1990). Modulation of fibroblast proliferation by oxygen free radicals. Biochem J 265: 659–665

Pender JR, Pories WJ (2005) Epidemiology of obesity in the United States. Gastroenterol Clin North Am 34: 1–7

Radak Z, Chung HY, Koltai E, Taylor AW, Goto S (2008) Exercise, oxidative stress and hormesis. Ageing Res Rev 7: 34–42

Ramirez-Ortega M, Maldonado-Lagunas V, Melendez-Zajgla J, Carrillo-Hernandez JF, Pastelin-Hernandez G, Picazo-Picazo O, Ceballos-Reyes G (2006). Proliferation and apoptosis of HeLa cells induces by in vitro stimulation with digitalis. Eur J Pharmacol 534: 71–86

Randall WA, Price CW, Welch H (1947). Demonstration of hormesis (increase in fatality rate) by penicillin. Am J Public Health 37: 421–425

Redman LM, Martin CK, Williamson DA, Ravussin E (2008) Effect of caloric restriction in non-obese humans on physiological, psychological and behavioral outcomes. Physiol Behav 94: 643–648

Smith-Sonneborn J (2008). Hormetic triggers for intervention in aging, disease and trauma. Am J Pharmacol Toxicol 3: 1–10

Strobel G, Daisy B, Castillo U, Harper J (2004) Natural products from endophytic microorganisms. J Nat Prod 67: 257–268

Triggiani V, Resta F, Guastamacchia E, Sabbà C, Licchelli B, Ghiyasaldin S, Tafaro E (2006) Role of antioxidants, essential fatty acids, carnitine, vitamins, phytochemicals and trace elements in the treatment of diabetes mellitus and its chronic complications. Endocr Metab Immune Disord Drug Targets 6: 77–93

van Praag H (2008) Neurogenesis and exercise: past and future directions. Neuromolecular Med 10: 128–140

van Praag H, Lucero MJ, Yeo GW, Stecker K, Heivand N, Zhao C, Yip E, Afanador M, Schroeter H, Hammerstone J, Gage FH (2007) Plant-derived flavanol (-)epicatechin enhances angiogenesis and retention of spatial memory in mice. J Neurosci 27: 5869–5878

Welch H, Price CW, Randall, WA (1946). Increase in fatality rate of *E. typhosa* for white mice by streptomycin. J Am Pharmaceut Assoc 35: 155–158

Yu ZF, Mattson MP (1999) Dietary restriction and 2-deoxyglucose administration reduce focal ischemic brain damage and improve behavioral outcome: evidence for a preconditioning mechanism. J Neurosci Res 57: 830–839

Index

M.P. Mattson, E.J. Calabrese *Hormesis*, DOI 10.1007/978-1-60761-495-1,

L

M